Frontiers in Mathematics

Yuming Qin
Lan Huang

Global
Well-posedness
of Nonlinear
Parabolic-Hyperbolic
Coupled
Systems

 Birkhäuser

Yuming Qin
Department of Applied Mathematics
Donghua University
Shanghai
People's Republic of China

Lan Huang
College of Mathematics and Information Science
North China University of Water Sources
and Electric Power
Zhengzhou
People's Republic of China

ISSN 1660-8046   e-ISSN 1660-8054
ISBN 978-3-0348-0279-6   e-ISBN 978-3-0348-0280-2
DOI 10.1007/978-3-0348-0280-2
Springer Basel Dordrecht Heidelberg London New York

Library of Congress Control Number: 2012931854

Mathematics Subject Classification (2010): 35Q30, 76-XX, 76D05

Printed on acid-free paper

Springer Basel AG is part of Springer Science+Business Media

www.birkhauser-science.com

To our parents

Zhenrong Qin   Xilan Xia

and

Shaolin Huang   Chuanfeng Yang

Contents

Preface ix

1 Global Existence of Spherically Symmetric Solutions for Nonlinear Compressible Non-autonomous Navier-Stokes Equations

  1.1   Introduction   1

  1.2   Global Existence of Solutions in H^1   4

  1.3   Global Existence of Solutions in H^2   19

  1.4   Global Existence of Solutions in H^4   25

  1.5   Bibliographic Comments   32

2 Global Existence and Exponential Stability for a Real Viscous Heat-conducting Flow with Shear Viscosity

  2.1   Introduction   33

  2.2   Proof of Theorem 2.1.1   38

  2.3   Proof of Theorem 2.1.2   46

  2.4   Proof of Theorem 2.1.3   53

  2.5   Bibliographic Comments   73

3 Regularity and Exponential Stability of the pth Power Newtonian Fluid in One Space Dimension

  3.1   Introduction   75

  3.2   Proof of Theorem 3.1.1   78

  3.3   Proof of Theorem 3.1.2   89

  3.4   Bibliographic Comments   91

4 Global Existence and Exponential Stability for the pth Power Viscous Reactive Gas

  4.1   Introduction   93

  4.2   Global Existence in H^2   96

  4.3   Exponential Stability in H^2   105

  4.4   Proof of Theorem 4.1.2   112

  4.5   Bibliographic Comments   125

5 On a 1D Viscous Reactive and Radiative Gas with
First-order Arrhenius Kinetics

5.1	Introduction	127
5.2	Global Existence in H^1	132
5.3	Exponential Stability in H^1	156
5.4	Proof of Theorem 5.1.2	157
5.5	Proof of Theorem 5.1.3	159
5.6	Bibliographic Comments	161

Bibliography 165

Index 171

Preface

This book aims to present some recent results on a selection of nonlinear parabolic-hyperbolic coupled systems arising from physics, fluid mechanics and material science such as the compressible Navier-Stokes equations, viscous reactive and radiative gas, etc. Most of the material in this book is based on research carried out by the authors and their collaborators in recent years. Some of it has been previously published only in original papers, and some of the material has never been published until now.

There are five chapters in this book. Chapters 1–2 will be concerned with the study of non-isentropic Navier-Stokes equations. In Chapter 1, we shall investigate the global existence of spherically symmetric solutions for nonlinear compressible Navier-Stokes equations of initial boundary value problems with a non-autonomous external force and a heat source in bounded annular domains $G_n = \{r \in \mathbb{R}^n : a \leq r \leq b\}$ in $\mathbb{R}^n$ ($1 \leq n \leq 3$). Thus we stress mainly the affect of non-autonomous terms on the global well-posedness and on the large-time behavior of solutions. Chapter 2 is concerned with the global existence and exponential stability of solutions for a real viscous compressible heat-conducting flow with general constitutive relations between two horizontal plates.

Chapters 3–4 are devoted to an investigation of the global existence and exponential stability of solutions to a pth power viscous reactive gas (Newtonian fluid) without and with a chemical reaction respectively. We should note that these models are quite different from the usual viscous heat-conductive reactive gas in Qin [59]. In detail, the models considered in Chapters 3–4 have more complicated constitutive relationships than those in Qin [59], and so we need to design more delicate and careful estimates to establish the uniform lower and upper bounds of the specific volume.

Chapter 5 is devoted to the study of viscous reactive and radiative gas. In Chapter 5, we shall establish the global existence and exponential stability of solutions in H^i ($i = 1, 2, 4$) for a Stefan-Boltzmann model of a viscous, reactive and radiative gas with first-order Arrhenius kinetics in a bounded interval, which describes classical stellar evolution of a finite mass of a heat-conducting viscous reactive fluid in local equilibrium with thermal radiation: pressure, internal energy and thermal conductivity have Stefan-Boltzmann radiative contributions.

We sincerely hope that the reader will learn the main ideas and essence of the basic theories and methods in deriving global well-posedness, asymptotic behavior and regularity of global solutions for the models under consideration in this book. Also we are confident that the reader will be stimulated by some ideas from this book and undertake further study and research after having read the related references and our bibliographic comments.

We extend our appreciation to Dr. Thomas Hempfling and Dr. Barbara Hellriegel of Birkhäuser/Springer Basel for their great efforts in publishing this book.

We also want to take this opportunity to thank all the people who supported us including our teachers, colleagues and collaborators, etc. Yuming Qin appreciates his former Ph. D. advisor, Professor Songmu Zheng, for his constant encouragement, great help and support. Also Qin thanks Professors Ta-tsien Li, Boling Guo, Jiaxing Hong, Yiming Long, Weixi Shen, Ling Hsiao, Shuxing Chen, Zhenting Hou, Long-an Ying, Guangjun Yang, Guowang Chen, Jinhua Wang, Junning Zhao, Tiehu Qin, Yongji Tan, Sining Zheng, Zhouping Xin, Tong Yang, Hua Chen, Chaojiang Xu, Jingxue Yin, Liqun Zhang, Weike Wang, Mingxin Wang, Fahuai Yi, Song Jiang, Chengkui Zhong, Xuguang Lu, Yinbin Deng, Daomin Cao, Cheng He, Yi Zhou, Xiangyu Zhou, Quansui Wu, Daoyuan Fang, Ping Zhang, Changjiang Zhu, Changxing Miao, Yongsheng Li, Feimin Huang, Huijiang Zhao, Zheng-an Yao, Changzheng Qu, Yaping Wu, Zhaoli Liu, Huicheng Yin, Xiaozhou Yang, Shu Wang, Yaguang Wang, Zhong Tan, Xingbin Pan, Feng Zhou, Baojun Bian, Shengliang Pan, Wen-an Yong, Boxiang Wang, Lixin Tian, Shangbin Cui, Shijin Ding, Xi-nan Ma, Huaiyu Jian, Yachun Li, Benjin Xuan, Ting Wei, Quansen Jiu, Hailiang Li, Jiabao Su, Kaijun Zhang, Peidong Lei, Yongqian Zhang, Zhaoyang Yin, Wenyi Chen, Zhigui Lin, Xiaochun Liu, Hao Wu, Ting Zhang, Zhenhua Guo, Hongjun Yu, Ning Jiang and Yawei Wei for their constant help. Also Qin would like to thank Professors Herbert Amann, Michel Chipot from Switzerland, Professors J.A. Burns, Taiping Liu, Guiqiang Chen, D. Gilliams, Irena Lasiecka, Joel Spruck, M. Slemrod, Yisong Yang, Zhuangyi Liu, T.H. Otway, Shouhong Wang, Yuxi Zheng, Chun Liu, Changfeng Gui, Shouchuan Hu, Jianguo Liu, Hailiang Liu, Tao Luo from the USA, Professors Roger Temam, Alain Miranville, D. Hilhorst, Vilmos Komornik, Mokhtar Kirane, Patrick Martinez, Fatiha Alabau-Boussouira, Yuejun Peng and Bopeng Rao from France, Professors Hugo Beirao da Veiga, Maurizio Grasselli, Cecilia Cavaterra from Italy, Professors Bert-Wolfgang Schulze, Reinhard Racke, Michael Reissig, Ingo Witt, Jürgen Sprekels, H.-D. Alber from Germany, Professors Enrique Zuazua, Peicheng Zhu from Spain, Professors Jaime E. Muñoz Rivera, Tofu Ma, Alexandre L. Madureira, Jinyun Yuan, D. Andrade, M.M. Cavalcanti, Frédéric G. Christian Valentin from Brazil, Professors Tzon Tzer Lü, Jyh-Hao Lee, Chun-Kong Law, Ngai-Ching Wong from Chinese Taiwan for their constant help.

Yuming Qin also acknowledges the NNSF of China for its support. Currently, this book project is being supported by the NNSF of China with contract no. 11031003 and no. 10871040 and by the Sino-German cooperation grant "Analysis of partial differential equations and applications" with contract no. 446 CHV 113/267/0-1.

Last but not least, Yuming Qin wants to express his deepest thanks to his parents (Zhenrong Qin and Xilan Xia), sisters (Yujuan Qin and Yuzhou Qin), brother (Yuxing Qin), wife (Yu Yin) and son (Jia Qin) for their great help, constant concern and advice in his career, and Lan Huang takes this opportunity to express thanks to her husband (Fengxiao Zhai) for his great support in her career.

Yuming Qin and Lan Huang

Chapter 1

Global Existence of Spherically Symmetric Solutions for Nonlinear Compressible Non-autonomous Navier-Stokes Equations

1.1 Introduction

This chapter concerns the global existence of spherically symmetric solutions for nonlinear compressible non-autonomous Navier-Stokes equations of an initial boundary value problem with an external force and a heat source in bounded annular domains $G_n = \{r \in \mathbb{R}^n : a \le r \le b\}$ in $\mathbb{R}^n$ $(1 \le n \le 3)$. In Eulerian coordinates, equations under consideration are expressed as:

$$\partial_t \rho + \partial_r(\rho v) + \frac{(n-1)}{r}\rho v = 0, \tag{1.1.1}$$

$$C_v \rho(\partial_t v + v\partial_r v) - (\lambda + 2\mu)\left[\partial_r^2 v + \frac{(n-1)}{r}\partial_r v - \frac{(n-1)}{r^2}v\right]$$
$$+ R\partial_r(\rho v) = f(r,t), \tag{1.1.2}$$

$$C_v \rho(\partial_t \theta + v\partial_r \theta) - \kappa\partial_r^2\theta - k\frac{(n-1)}{r}\partial_r\theta + R\rho\theta\left[\partial_r v + \frac{(n-1)}{r}v\right]$$
$$- \lambda\left[\partial_r v + \frac{(n-1)}{r}v\right]^2 - 2\mu(\partial_r v)^2 - 2\mu\frac{(n-1)}{r^2}v^2 = g(r,t). \tag{1.1.3}$$

We consider equations (1.1.1)–(1.1.3) with initial boundary value conditions

$$\rho(r,0) = \rho_0(r), \quad v(r,0) = v_0(r), \quad \theta(r,0) = \theta_0(r), \quad r \in G_n, 1 \le n \le 3, \tag{1.1.4}$$

$$v(a,t) = v(b,t) = 0, \quad \theta_r(a,t) = \theta_r(b,t) = 0, 1 \le n \le 3 \tag{1.1.5}$$

where constants $R, C_v, \kappa, \mu > 0$ and $\beta = \lambda + 2\mu$.

Equations (1.1.1)–(1.1.3) describe the symmetric motion of a viscous poly-tropic ideal gas in n-dimension with an external force f and a heat source g where ρ, v, θ are the density, velocity, and absolute temperature, λ and μ are the constant viscosity coefficients, R, C_v, and κ are the gas constant, specific heat capacity and thermal conductivity, respectively.

It is convenient to transform system (1.1.1)–(1.1.3) into Lagrangian coordinates. The Eulerian coordinates (r, t) are connected to the Lagrangian coordinates (ζ, t) by the relation

$$r(\zeta, t) = r_0(\zeta) + \int_0^t \tilde{v}(\zeta, \tau)d\tau, \tag{1.1.6}$$

where

$$\tilde{v}(\zeta, t) := v(r(\zeta, t), t), \quad r_0(\zeta) := \eta^{-1}(\zeta), \quad \eta(r) := \int_{d_n}^r s^{n-1}\rho_0(s)ds,$$

$$r \in G_n, \quad d_n = 0, \ (n = 1), \quad d_n = a, \ (n = 2, 3).$$

Suppose that $\rho_0(s) > 0, s \in G_n$. Set $L = \int_a^b s^{n-1}\rho_0(s)ds > 0$. Using the equation (1.1.1), (1.1.5) and (1.1.6), we obtain

$$\partial_t \int_{d_n}^{r(\zeta,t)} s^{n-1}\rho(s, t)ds = \delta_{n1}v(0, t)\rho(0, t), \quad \delta_{ij} = 1, (i = j); \quad \delta_{ij} = 0, (i \neq j).$$

By integration, we derive

$$\int_{d_n}^{r(\zeta,t)} s^{n-1}\rho(s, t)ds = \int_{d_n}^{r_0(\zeta)} s^{n-1}\rho_0(s)ds + \delta_{n1}\int_0^t (v\rho)(0, \tau)d\tau$$

$$= \zeta + \delta_{n1}\int_0^t (v\rho)(0, \tau)d\tau.$$

Thus, under the assumption $\inf\{\rho(s, t); s \in G_n, t \geq 0\} > 0$, G_n is transformed to Ω_n, with $\Omega_n = (0, L)$, $(n = 1, 2, 3)$. Moreover, we have

$$\partial_\zeta r(\zeta, t) = [r(\zeta, t)^{n-1}\rho(r(\zeta, t), t)]^{-1}. \tag{1.1.7}$$

For a function $\varphi(r, t)$, we write $\tilde{\varphi}(\zeta, t) := \varphi(r(\zeta, t), t)$. By virtue of (1.1.6) and (1.1.7), we derive

$$\partial_t \tilde{\varphi}(\zeta, t) = \partial_t \varphi(r, t) + v\partial_r \varphi(r, t),$$

$$\partial_\zeta \tilde{\varphi}(\zeta, t) = \partial_r \varphi(r, t)\partial_\zeta r(\zeta, t) = \frac{1}{r^{n-1}\rho(r, t)}\partial_r \varphi(r, t). \tag{1.1.8}$$

We still denote $(\tilde{\rho}, \tilde{v}, \tilde{\theta})$ by (ρ, v, θ) and (ζ, t) by (x, t). We use $u := \frac{1}{\rho}$ to denote the specific volume. Therefore, by virtue of (1.1.7)–(1.1.8), equations (1.1.1)–(1.1.5) in the new variables (x, t) are

$$u_t - (r^{n-1}v)_x = 0, \tag{1.1.9}$$

$$v_t - r^{n-1}\left(\beta\frac{(r^{n-1}v)_x}{u} - R\frac{\theta}{u}\right)_x = f(r(x,t),t), \tag{1.1.10}$$

$$C_v\theta_t - k\left(\frac{r^{2n-2}\theta_x}{u}\right)_x - \frac{1}{u}(\beta(r^{n-1}v)_x - R\theta)(r^{n-1}v)_x + 2\mu(n-1)(r^{n-2}v^2)_x$$
$$= g(r(x,t),t), \tag{1.1.11}$$

together with

$$u(x,0) = u_0(x), \quad v(x,0) = v_0(x), \quad \theta(x,0) = \theta_0(x), \quad x \in \Omega_n, \ 1 \le n \le 3, \tag{1.1.12}$$

$$v(0,t) = v(L,t) = 0, \quad \theta_x(0,t) = \theta_x(L,t) = 0, \quad t \ge 0, \ 1 \le n \le 3 \tag{1.1.13}$$

where $\beta = \lambda + 2\mu$. Then by (1.1.6), we have

$$r(x,t) = r_0(x) + \int_0^t v(x,\tau)d\tau, \quad r_0(x) := \left\{(d_n)^n + n\int_0^x u_0(y)dy\right\}^{\frac{1}{n}},$$

i.e.,

$$r_t = v, \quad r^{n-1}r_x = u, \quad r|_{x=0} = a, \quad r|_{x=L} = b. \tag{1.1.14}$$

When $n = 2,3$, for constants λ, μ, we suppose

$$n\lambda + 2\mu > 0. \tag{1.1.15}$$

The aim of this chapter is to investigate the global existence of solutions in H^i $(i = 1,2,4)$. The global existence of solutions in H^4 implies the global existence of classical solutions. We finally obtain the results by the energy method as in [52, 53, 54, 55, 57, 58, 65, 67, 68, 69].

We suppose that $f(r,t)$, $g(r,t)$ satisfy:

$$f \in L^1(\mathbb{R}^+, L^\infty[a,b]) \cap L^2(\mathbb{R}^+, L^2[a,b]), \tag{1.1.16}$$
$$g \in L^1(\mathbb{R}^+, L^\infty[a,b]) \cap L^2(\mathbb{R}^+, L^2[a,b]),$$
$$g(r,t) \ge 0, \forall(r,t) \in [a,b] \times [0,+\infty). \tag{1.1.17}$$

The notation of this section is as follows: $\|\cdot\|_B$ denotes the norm of space B, $\|\cdot\| = \|\cdot\|_{L^2}$. C_1 denotes a generic positive constant depending only on the H^1 norm of initial data (u_0, v_0, θ_0), $\|f\|_{L^1(\mathbb{R}^+, L^\infty[a,b])}$, $\|f\|_{L^2(\mathbb{R}^+, L^2[a,b])}$, $\|g\|_{L^1(\mathbb{R}^+, L^\infty[a,b])}$, $\|g\|_{L^2(\mathbb{R}^+, L^2[a,b])}$, but is independent of t. The results of this chapter are selected from [72].

1.2   Global Existence of Solutions in H^1

In this section, we shall establish the global existence of solutions in H^1.

Theorem 1.2.1. *Assume that* (1.1.16)–(1.1.17) *hold. If* $(u_0, v_0, \theta_0) \in H^1[0, L] \times H_0^1[0, L] \times H^1[0, L]$, *then problem* (1.1.9)–(1.1.15) *admits a unique global solution* $(u(t), v(t), \theta(t)) \in C([0, +\infty), H^1[0, L] \times H_0^1[0, L] \times H^1[0, L])$ *satisfying*

$$0 < a \le r(x, t) \le b, \quad (x, t) \in [0, L] \times [0, +\infty), \tag{1.2.1}$$

$$0 < C_1^{-1} \le u(x, t) \le C_1, \quad (x, t) \in [0, L] \times [0, +\infty), \tag{1.2.2}$$

$$\|r(t) - \bar{r}\|_{H^2}^2 + \|r_t(t)\|_{H^1}^2 + \|u(t) - \bar{u}\|_{H^1}^2 + \|v(t)\|_{H^1}^2 + \|\theta(t) - \bar{\theta}\|_{H^1}^2$$
$$\quad + \|u(t) - \bar{u}\|_{L^\infty}^2 + \|v(t)\|_{L^\infty}^2 + \|\theta(t) - \bar{\theta}\|_{L^\infty}^2 + \|u_t(t)\|^2$$
$$\quad + \int_0^t \left(\|u - \bar{u}\|_{H^1}^2 + \|v\|_{H^2}^2 + \|\theta - \bar{\theta}\|_{H^2}^2 + \|u_t\|_{H^1}^2 + \|v_t\|^2 \right.$$
$$\quad \left. + \|\theta_t\|^2 + \|r - \bar{r}\|_{H^2}^2 + \|r_t\|_{H^2}^2 \right)(\tau)d\tau \le C_1, \quad \forall t \ge 0, \tag{1.2.3}$$

where

$$\bar{u} = \frac{1}{L} \int_0^L u_0(x)dx, \qquad \bar{r} = (a^n + n\bar{u}x)^{\frac{1}{n}}, \qquad \bar{\theta} = \frac{1}{C_v L} \int_0^L \left(C_v \theta_0 + \frac{v_0^2}{2} \right)(x)dx.$$

Lemma 1.2.1.

$$a = r(0, t) \le r(x, t) \le r(L, t) = b, \quad \forall x \in [0, L], \quad \forall t \ge 0. \tag{1.2.4}$$

Proof. By (1.1.14), it holds that $r_x(0, t) = r^{1-n}(0, t)u(0, t) = a^{1-n}u(0, t) > 0$, $\forall t \ge 0$. Suppose that $r_x(x, t)$ is not always positive over $[0, L] \times [0, +\infty)$, i.e., there are $y \in (0, L]$, $\tau \in [0, +\infty)$, such that $r_x(x, t) > 0$ over $[0, y) \times [0, \tau]$, but $r_x(y, \tau) = 0$. So, by continuity, $r_x(x, t) \ge 0$ over $[0, y] \times [0, \tau]$, therefore $r(y, \tau) \ge r(0, \tau) = a > 0$. From (1.1.14), $0 = r_x(y, \tau) = r^{1-n}(y, \tau)u(y, \tau) > 0$, which is a contradiction. This shows that $r_x(x, t) > 0$ over $[0, L] \times [0, +\infty)$. Hence, $a = r(0, t) \le r(x, t) \le r(L, t) = b$, $\quad \forall x \in [0, L]$, $\quad \forall t > 0$. □

Remark 1.2.1. It follows from Lemma 1.2.1 that assumptions (1.1.16) and (1.1.17) are equivalent to the following conditions: $\forall x \in [0, L]$,

$$f(r(x, t), t) \in L^1(\mathbb{R}^+, L^\infty[a, b]) \cap L^2(\mathbb{R}^+, L^2[a, b]), \tag{1.2.5}$$

$$g(r(x, t), t) \in L^1(\mathbb{R}^+, L^\infty[a, b]) \cap L^2(\mathbb{R}^+, L^2[a, b]),$$
$$g(r(x, t), t) \ge 0, \; \forall(x, t) \in [0, L] \times [0, +\infty). \tag{1.2.6}$$

Thus the generic positive constant C_1 depends only on the H^1 norm of initial data (u_0, v_0, θ_0) and norms of f and g in the class of functions in (1.2.5)–(1.2.6), but independent of t.

Lemma 1.2.2. *Under the conditions of Theorem 1.2.1, there is a constant $C_1 > 0$ such that*

$$\int_0^L U(x,t)dx + \int_0^t \int_0^L \left(\frac{v_x^2}{u\theta} + \frac{\theta_x^2}{u\theta^2} + \frac{g}{\theta} \right)(x,s)dxds \leq C_1, \quad \forall t > 0, \quad (1.2.7)$$

where

$$U(x,t) \triangleq \left\{ \frac{v^2}{2} + R(u - \ln u - 1) + C_v(\theta - \ln\theta - 1) \right\}(x,t). \quad (1.2.8)$$

Proof. By equations (1.1.9)–(1.1.11), we can easily obtain

$$U_t + \frac{\beta}{u\theta}(r^{n-1}v)_x^2 + \frac{\kappa}{u\theta^2}(r^{n-1}\theta_x)^2$$

$$= \left\{ r^{n-1}v \left(\frac{\beta}{u}(r^{n-1}v)_x - R\frac{\theta}{u} \right) \right\}_x + vf + R(r^{n-1}v)_x + \kappa \left\{ \left(1 - \frac{1}{\theta} \right) \frac{r^{2n-2}\theta_x}{u} \right\}_x$$

$$- 2(n-1)\mu \left(1 - \frac{1}{\theta} \right)(r^{n-2}v^2)_x + g - \frac{g}{\theta}. \quad (1.2.9)$$

By (1.1.14) and (1.1.15), we easily derive

$$\frac{\beta}{u\theta}\left(r^{n-1}v\right)_x^2 - 2\mu(n-1)\frac{(r^{n-2}v^2)_x}{\theta}$$

$$= \frac{1}{u\theta}\left\{ (n-1)(2\mu + (n-1)\lambda) \left(r^{-1}uv + \frac{\lambda r^{n-1}v_x}{2\mu + (n-1)\lambda} \right)^2 \right.$$

$$\left. + \frac{2\mu(2\mu + n\lambda)}{2\mu + (n-1)\lambda}r^{2n-2}v_x^2 \right\}$$

$$\geq \frac{2\mu(2\mu + n\lambda)}{2\mu + (n-1)\lambda}\frac{r^{2n-2}v_x^2}{u\theta}. \quad (1.2.10)$$

Integrating (1.2.9) over $[0, L] \times [0, t]$, noting that

$$\int_0^L U(x,0)dx \leq C_1 \left(1 + \|(u_0, v_0, \theta_0)\|^2 \right) \leq C_1,$$

and considering (1.1.12), (1.2.10), we can obtain

$$\int_0^L U(x,t)dx + \int_0^t \int_0^L \left(\frac{v_x^2}{u\theta} + \frac{\theta_x^2}{u\theta^2} + \frac{g}{\theta} \right)(x,s)dxds$$

$$\leq C_1 + \int_0^t \int_0^L (vf + g)(x,s)dxds \quad (1.2.11)$$

where

$$\left| \int_0^t \int_0^L vf\,dxds \right| \le C_1 \int_0^t \int_0^L |f| v^2\,dxds + C_1 \int_0^t \int_0^L |f|\,dxds$$

$$\le C_1 \int_0^t \sup_{x\in[0,L]} |f| \int_0^L v^2\,dxds + C_1 \int_0^t \int_0^L |f|\,dxds. \quad (1.2.12)$$

Combining (1.2.11)–(1.2.12), using Gronwall's inequality and conditions (1.1.16)–(1.1.17), we can obtain (1.2.2). $\square$

Lemma 1.2.3. *There are positive constants α_1 and α_2 such that*

$$\alpha_1 \le \int_0^L \theta(x,t)dx \le \alpha_2, \quad \forall t > 0, \quad (1.2.13)$$

and there is a point $a(t) \in [0,L]$ satisfying

$$\frac{\alpha_1}{L} \le \theta(a(t),t) \le \frac{\alpha_2}{L}. \quad (1.2.14)$$

Proof. From (1.2.7), we have

$$C_v \int_0^L \left(\theta(x,t) - \log\theta(x,t) - 1\right)(x,t)dx \le C_1, \quad \forall t > 0. \quad (1.2.15)$$

By virtue of the mean value theorem, for any $t \ge 0$, there is a point $a(t) \in [0,L]$ such that

$$\theta(a(t),t) - \log\theta(a(t),t) - 1 \le (C_v L)^{-1} C_1$$

from which it follows that $\zeta_1 \le \theta(a(t),t) \le \zeta_2$, with ζ_1, ζ_2 being two positive roots of the equation: $y - \ln y - 1 = C_v^{-1} C_1$. By Jensen's inequality, we can obtain

$$\int_0^L \theta(x,t)dx - \log\int_0^L \theta(x,t)dx - 1 \le C_v^{-1} C_1, \quad \forall t > 0.$$

Therefore $0 < \zeta_3 \le \int_0^L \theta(x,t)dx \le \zeta_4$, where ζ_3, ζ_4 are two positive roots of equation $y - \ln y - 1 = (C_v L)^{-1} C_1$. Taking $\alpha_1 = \min\{\zeta_1, \zeta_3\}$, $\alpha_2 = \max\{\zeta_2, \zeta_4\}$, we obtain (1.2.13) and (1.2.14). $\square$

Let

$$\sigma(x,t) := \beta\frac{(r^{n-1}v)_x}{u} - R\frac{\theta}{u}, \quad (1.2.16)$$

$$\phi(x,t) := \int_0^t \sigma(x,s)ds + \int_0^x r_0^{-(n-1)}(y)v_0(y)dy$$

$$+ (n-1)\int_0^t \int_x^L r^{-n}(y,s)v^2(y,s)dyds. \quad (1.2.17)$$

We deduce after a direct calculation that

$$\phi_x(x,t) = \int_0^t \sigma_x(x,s)ds + r_0^{1-n}(x)v_0(x) - (n-1)\int_0^t r^{-n}(x,s)v^2(x,s)ds$$

$$= r^{1-n}v(x,t) - \int_0^t r^{1-n}fds, \tag{1.2.18}$$

$$\phi_t(x,t) = \sigma(x,t) + (n-1)\int_x^L r^{-n}(y,t)v^2(y,t)dy$$

$$= \beta\frac{(r^{n-1}v)_x}{u} - R\frac{\theta}{u} + \frac{n-1}{n}\frac{(r^n)_x}{u}\int_x^L r^{-n}v^2dy. \tag{1.2.19}$$

Multiplying (1.2.19) by u, then from (1.1.9), (1.2.18), we derive

$$(u\phi)_t - (r^{n-1}v\phi)_x \tag{1.2.20}$$

$$= -\frac{v^2}{n} - R\theta + \beta(r^{n-1}v)_x + \frac{n-1}{n}(r^n)_x\int_x^L r^{-n}v^2dy + r^{n-1}v\int_0^t r^{1-n}fds$$

$$= -\frac{v^2}{n} - R\theta + \beta(r^{n-1}v)_x + \frac{n-1}{n}\left[r^n\int_x^L r^{-n}v^2dy\right]_x + r^{n-1}v\int_0^t r^{1-n}fds.$$

Integrating (1.2.20) over $[0,L]\times[0,t]$, we can get

$$\int_0^L u\phi dx = \int_0^L u_0(x)\phi_0(x)dx - \int_0^t\int_0^L\left(\frac{v^2}{n} + R\theta\right)dxds \tag{1.2.21}$$

$$- \frac{n-1}{n}a^n\int_0^t\int_0^L r^{-n}v^2dxds + \int_0^t\int_0^L r^{n-1}v\left(\int_0^s r^{1-n}fd\tau\right)dxds.$$

It follows from integration of (1.1.9) that

$$\int_0^L udx = \int_0^L u_0dx := u^*, \quad \forall t > 0. \tag{1.2.22}$$

Applying the mean value theorem, we infer that for any $t \geq 0$ there exists a point $x_0(t) \in [0,L]$ such that

$$\int_0^L \phi udx = \phi(x_0(t),t)\int_0^L udx = u^*\phi(x_0(t),t),$$

i.e.,

$$\phi(x_0(t),t) = \frac{1}{u^*}\int_0^L u\phi dx. \tag{1.2.23}$$

From (1.2.17), (1.2.21) and (1.2.23), we derive

$$\int_0^t \sigma(x_0(t), t)ds = \phi(x_0(t), t) - \int_0^{x_0(t)} r_0^{1-n} v_0 dy - (n-1) \int_0^t \int_{x_0(t)}^L r^{-n} v^2 dy ds$$

$$= -\frac{1}{u^*} \int_0^t \int_0^L \left(\frac{v^2}{n} + R\theta \right) dx ds - \frac{(n-1)a^n}{nu^*} \int_0^t \int_0^L r^{-n} v^2 dx ds$$

$$- (n-1) \int_0^t \int_{x_0(t)}^L r^{-n} v^2 dx ds + \frac{1}{u^*} \int_0^L u_0 \phi_0 dx - \int_0^{x_0(t)} r_0^{1-n} v_0 dy$$

$$+ \frac{1}{u^*} \int_0^t \int_0^L r^{n-1} v \left(\int_0^s r^{1-n} f d\tau \right) dx ds, \quad \forall t \ge 0. \tag{1.2.24}$$

In order to estimate the bounds of u, we first give a representation of u.

Lemma 1.2.4. *We have the following representation:*

$$u(x, t) = \frac{D(x, t)}{B(x, t)} \left[1 + \frac{R}{\beta} \int_0^t \frac{\theta(x, s)B(x, s)}{D(x, s)} ds \right], \quad x \in [0, L], \quad t > 0, \tag{1.2.25}$$

where

$$D(x, t) = u_0(x) \exp \left\{ \frac{1}{\beta} \left[\frac{1}{u^*} \int_0^L u_0 \phi_0 dx - \int_0^x r^{1-n} v_0 dy + \int_{x_0(t)}^x r^{1-n} v dy \right] \right.$$

$$+ \frac{1}{u^*} \int_0^t \int_0^L r^{n-1} v \left(\int_0^s r^{1-n} f d\tau \right) dx ds - \int_{x_0(t)}^x \int_0^t r^{1-n} f ds dy \right\}, \tag{1.2.26}$$

$$B(x, t) = \exp \left\{ \frac{1}{\beta} \left[\frac{1}{u^*} \int_0^t \int_0^L \left(\frac{v^2}{n} + R\theta \right) dx ds + \frac{(n-1)a^n}{nu^*} \int_0^t \int_0^L r^{-n} v^2 dx ds \right. \right.$$

$$+ (n-1) \int_0^t \int_x^L r^{-n} v^2 dy ds \right] \right\}. \tag{1.2.27}$$

Proof. By (1.1.9) and (1.1.10), we have

$$r^{1-n} v_t = \beta (\ln u)_{xt} - R \left(\frac{\theta}{u} \right)_x + r^{1-n} f. \tag{1.2.28}$$

Integrating (1.2.28) over $[0, t] \times [x_0(t), x]$, we can get

$$\beta \log u - R \int_0^t \frac{\theta}{u} ds = \beta \log u_0 + \int_0^t \sigma(x_0(s), s) ds + \int_{x_0(t)}^x \int_0^t r^{1-n} v_t ds dy$$

$$- \int_{x_0(t)}^x \int_0^t r^{1-n} f ds dx. \tag{1.2.29}$$

From (1.2.24) and (1.2.29), we have

$$\beta \log u - R \int_0^t \frac{\theta}{u} ds$$

$$= \beta \log u_0 - \frac{1}{u^*} \int_0^t \int_0^L \left(\frac{v^2}{n} + R\theta \right) dx ds - \frac{(n-1)a^n}{nu^*} \int_0^t \int_0^L r^{-n} v^2 dx ds$$

$$+ \frac{1}{u^*} \int_0^L u_0 \phi_0 dx - (n-1) \int_0^t \int_x^L r^{-n} v^2 dy ds - \int_0^x r_0^{1-n} v_0 dy + \int_{x_0}^x r^{1-n} v dy$$

$$- \int_{x_0(t)}^x \int_0^t r^{1-n} f ds dy + \frac{1}{u^*} \int_0^t \int_0^L r^{n-1} v \left(\int_0^s r^{1-n} f d\tau \right) dx ds.$$

With the definition of $B(x,t)$ and $D(x,t)$, we obtain

$$\frac{B(x,t)}{D(x,t)} = \frac{1}{u(x,.t)} \exp\left(\frac{R}{\beta} \int_0^t \frac{\theta(x,s)}{u(x,s)} ds \right). \tag{1.2.30}$$

Multiplying (1.2.30) by $\dfrac{R\theta}{\beta}$, then integrating the result with respect to t, we derive

$$\exp\left(\frac{R}{\beta} \int_0^t \frac{\theta(x,t)}{u(x,t)} ds \right) = 1 + \frac{R}{\beta} \int_0^t \frac{\theta(x,t)B(x,t)}{D(x,t)} ds. \tag{1.2.31}$$

Combining (1.2.31) and (1.2.30), we obtain (1.2.25). □

Lemma 1.2.5. *Under the conditions of Theorem 1.2.1, there are positive constants $\underline{u}$ and $\bar{u}$ such that*

$$\underline{u} \le u(x,t) \le \bar{u}, \quad \forall(x,t) \in [0,L] \times [0,+\infty). \tag{1.2.32}$$

Proof. We deduce after a direct calculation that

$$\left| \int_{x_0(t)}^x \int_0^t r^{1-n} f ds dy \right| \le a^{1-n} \int_0^t \int_0^L |f| dy ds \le C_1, \tag{1.2.33}$$

$$\left| \int_{x_0(t)}^x r^{1-n} v dy \right| \le a^{1-n} \int_0^L |v| dy \le a^{1-n} \|v\| \le C_1. \tag{1.2.34}$$

By Lemma 1.2.1, condition (1.1.16), and (1.2.12), we obtain

$$\left| \int_0^t \int_0^L r^{n-1} v \left(\int_0^s r^{1-n} f d\tau \right) dx ds \right| \le C_1 \int_0^t \int_0^L |v| \left(\int_0^\infty \sup_{x \in [0,L]} |f| d\tau \right) dx ds$$

$$\le C_1 \int_0^t \int_0^L v.\operatorname{sgn} v \, dx ds = C_1 \int_0^t \int_0^L (r \operatorname{sgn} v)_s \, dx ds$$

$$= C_1 \int_0^L (r \operatorname{sgn} v - r_0 \operatorname{sgn} v_0) \, dx \le C_1. \tag{1.2.35}$$

From (1.2.33)–(1.2.35), we obtain

$$0 < C_1^{-1} \le D(x,t) \le C_1, \quad \forall x \in [0,L], \quad \forall t > 0. \tag{1.2.36}$$

By the definition of $B(x,t)$, (1.2.1) and (1.2.13), we can get

$$\frac{B(x,s)}{B(x,t)} = \exp\left\{ -\frac{R}{\beta u^*} \int_s^t \int_0^L \theta(x,s)dxds \right\} \le \exp\left\{ -\frac{R\alpha_1(t-s)}{\beta u^*} \right\}, \quad (t \ge s \ge 0).$$

Similarly,

$$\frac{B(x,s)}{B(x,t)} \ge C_1 e^{-C_1(t-s)}, \quad t \ge s \ge 0. \tag{1.2.37}$$

Therefore, it follows from Lemmas 1.2.1–1.2.3 that

$$1 \le B(x,t) \le e^{C_1 t}, \quad \forall t > 0. \tag{1.2.38}$$

By the Hölder inequality and (1.2.13), we have

$$| \theta^{\frac{1}{2}}(x,t) - \theta^{\frac{1}{2}}(a(t),t) | \le C_1 \int_0^L \theta^{-\frac{1}{2}}(x,t) \, | \, \theta_x(x,t) \, | \, dx$$

$$\le C_1 \left\{ \int_0^L \frac{\theta_x^2}{u\theta^2}(x,t)dx \right\}^{\frac{1}{2}} \left(\int_0^L \theta u dx \right)^{\frac{1}{2}}$$

$$\le C_1 \left\{ \int_0^L \frac{\theta_x^2}{u\theta^2}(x,t)dx \right\}^{\frac{1}{2}} \sup_{x \in [0,L]} u^{\frac{1}{2}}(x,t)$$

which, together with (1.2.13), implies

$$\frac{\alpha_1}{2} - C_1^{-1} \sup_{x \in [0,L]} u(\cdot,t) \int_0^L \frac{\theta_x^2}{u\theta^2}dx \le \theta(x,t)$$

$$\le 2\alpha_2 + C_1 \sup_{x \in [0,L]} u(\cdot,t) \int_0^L \frac{\theta_x^2}{u\theta^2}dx, \quad \forall(x,t) \in [0,L] \times [0,+\infty). \tag{1.2.39}$$

From (1.2.25), (1.2.36) and (1.2.38), we derive

$$u(x,t) \le C_1 + C_1 \int_0^t \left\{ 1 + \sup_{x \in [0,L]} u(\cdot,s) \int_0^L \frac{\theta_x^2}{u\theta^2}(x,s)dx \right\} e^{-\frac{t-s}{L}} ds$$

$$\le C_1 + C_1 \int_0^t \sup_{x \in [0,L]} u(\cdot,s) \int_0^L \frac{\theta_x^2}{u\theta^2}(x,s)dxds.$$

By (1.2.8) and Gronwall's inequality, we obtain

$$u(x,t) \le \bar{u}, \quad \forall(x,t) \in [0,L] \times [0,+\infty).$$

From (1.2.7), (1.2.25), (1.2.36), (1.2.38) and (1.2.39), we have

$$
u(x,t) \geq \frac{RD(x,t)}{\beta} \int_0^t \frac{\theta(x,s)B(x,s)}{D(x,s)B(x,t)} ds
$$

$$
\geq C_1 \int_0^t \left\{ \frac{\alpha_1}{2} - C_1^{-1} \sup_{x \in [0,L]} u(\cdot, s) \int_0^L \frac{\theta_x^2}{u\theta^2}(x,s)dx \right\} e^{-C_1(t-s)} ds
$$

$$
\geq \frac{\alpha_1}{2}(1 - e^{-C_1 t}) - C_1^{-1} e^{-\frac{C_1 t}{2}} \int_0^{+\infty} \int_0^L \frac{\theta_x^2}{u\theta^2}(x,s)dxds
$$

$$
- C_1^{-1} \int_{\frac{t}{2}}^t \int_0^L \frac{\theta_x^2}{u\theta^2}(x,s)dxds.
$$

Note that

$$
\int_{\frac{t}{2}}^t \int_0^L \frac{\theta_x^2}{u\theta^2}(x,s)dxds \longrightarrow 0 \quad \text{as} \quad t \longrightarrow +\infty;
$$

therefore there is a sufficiently large time $T_0 > 0$ such that when $t > T_0$, $u(x,t) \geq \frac{\alpha_1}{4} > 0$. When $t \in [0, T_0]$,

$$
u(x,t) \geq \frac{D(x,t)}{B(x,t)} \geq C_1^{-1} e^{-C_1 t} \geq C_1^{-1} e^{-C_1 T_0}. \tag{1.2.40}
$$

Taking $\underline{u} = \min\left\{ \frac{\alpha_1}{4}, C_1^{-1} e^{-C_1 T_0} \right\} > 0$, we obtain (1.2.32). $\square$

Lemma 1.2.6. *Under the conditions of Theorem 1.2.1, we have*

$$
\int_0^L (\theta^2 + v^4)(x,t)dx + \int_0^t \int_0^L (v^2 v_x^2 + \theta_x^2)(x,s)dxds \leq C_1, \quad \forall t > 0. \tag{1.2.41}
$$

Proof. From equations (1.1.9)–(1.1.11), we obtain

$$
\left(C_v \theta + \frac{v^2}{2} \right)_t
$$

$$
= k \left(\frac{r^{2n-2}\theta_x}{u} \right)_x + \frac{1}{u} \left[\beta(r^{n-1}v)_x - R\theta \right] (r^{n-1}v)_x - 2\mu(n-1)(r^{n-2}v^2)_x
$$

$$
+ g + r^{n-1}v \left[\beta\frac{(r^{n-1}v)_x}{u} - R\frac{\theta}{u} \right]_x + vf
$$

$$
= \left[k \left(\frac{r^{2n-2}\theta_x}{u} \right) + (r^{n-1}v) \left(\frac{\beta(r^{n-1}v)_x}{u} - R\frac{\theta}{u} \right) - 2\mu(n-1)(r^{n-2}v^2)) \right]_x
$$

$$
+ g + vf. \tag{1.2.42}
$$

Multiplying (1.2.42) by $C_v\theta + \dfrac{v^2}{2}$, then integrating the result over $[0,L] \times [0,t]$, we have

$$\frac{1}{2}\int_0^L \left(C_v\theta + \frac{v^2}{2}\right)^2 (x,t)dx$$

$$\leq C_1 - \frac{C_v k}{2}\int_0^t \int_0^L \frac{r^{2n-2}\theta_x^2}{u}dxds + C_1\int_0^t\int_0^L (r^{2n-2}v_x^2v^2 + v^4 + \theta^2v^2)dxds$$

$$+ \int_0^t\int_0^L \left(C_v\theta + \frac{v^2}{2}\right)(g + vf)dxds, \qquad (1.2.43)$$

where

$$\int_0^t\int_0^L \left(C_v\theta + \frac{v^2}{2}\right)gdxds \leq \int_0^t \|g\|_{L^\infty[0,L]}\int_0^L \left(C_v\theta + \frac{v^2}{2}\right)dxds \leq C_1,$$

$$(1.2.44)$$

$$\left|\int_0^t\int_0^L \left(C_v\theta + \frac{v^2}{2}\right)vfdxds\right| \leq \int_0^t\int_0^L (\theta^2v^2 + f^2)dxds$$

$$+ C\int_0^t \sup_{x\in[0,L]}|f|\int_0^L (v^4 + v^2)dxds. \qquad (1.2.45)$$

Multiplying (1.1.10) by v^3, then integrating the result over $[0,L] \times [0,t]$, we get

$$\int_0^L\int_0^t v^3 v_t dsdx = \int_0^t\int_0^L r^{n-1}\left(\beta\frac{(r^{n-1}v)_x}{u} - R\frac{\theta}{u}\right)_x v^3 dxds + \int_0^t\int_0^L v^3 f dxds.$$

Integrating by parts and using the mean value theorem, we obtain

$$\frac{1}{4}\int_0^L v^4 dx \leq C_1 - C_1\int_0^t\int_0^L v^2 v_x^2 dxds + C_1\int_0^t\int_0^L \theta^2 v^2 dxds$$

$$+ C_1\int_0^t \sup_{x\in[0,L]}|f|\int_0^L (v^4 + v^2)dxds$$

$$\leq C_1 - C_1\int_0^t\int_0^L v^2 v_x^2 dxds + C_1\int_0^t\int_0^L \theta^2 v^2 dxds$$

$$+ C_1\int_0^t \sup_{x\in[0,L]}|f|\int_0^L v^4 dxds. \qquad (1.2.46)$$

Combining (1.2.43)–(1.2.46), we can get

$$\int_0^L (\theta^2 + v^4)dx + \int_0^t\int_0^L (v^2 v_x^2 + \theta_x^2)dxds \qquad (1.2.47)$$

$$\leq C_1 + C_1\int_0^t \left(\sup_{x\in[0,L]} v^2 + \sup_{x\in[0,L]}|f|\right)\int_0^L (\theta^2 + v^4)dxds + C_1\int_0^t\int_0^L v^4 dxds.$$

By Gronwall's inequality and noting that

$$\int_0^t \sup_{x\in[0,L]} v^2 ds \leq \int_0^t \left(\int_0^L |v_x| dx\right)^2 ds \leq C_1 \int_0^t \int_0^L \frac{v_x^2}{u\theta} dx \int_0^L u\theta dx ds \leq C_1,$$

$$\int_0^t \int_0^L v^4 dx ds \leq \int_0^t \sup_{x\in[0,L]} v^2 \left(\int_0^L v^2 dx\right) ds \leq C_1,$$

we can obtain (1.2.41).   □

Lemma 1.2.7. *Under the conditions of Theorem 1.2.1, we have*

$$\int_0^L u_x^2(x,t) dx + \int_0^t \int_0^L (v_x^2 + u_x^2 + \theta u_x^2)(x,s) dx ds \leq C_1, \quad \forall t > 0. \quad (1.2.48)$$

Proof. Multiplying (1.1.10) by v, then integrating the result over $[0,L] \times [0,t]$, we can get

$$\frac{1}{2}\int_0^L v^2 dx = \frac{1}{2}\int_0^L v_0^2 dx - \int_0^t \int_0^L \beta \frac{(r^{n-1}v)_x^2}{u} dx ds$$

$$- \int_0^t \int_0^L R\left(\frac{\theta}{u}\right)_x (r^{n-1}v) dx ds + \int_0^t \int_0^L vf dx ds.$$

By Young's inequality and the embedding theorem, we have for any $\varepsilon > 0$,

$$\frac{1}{2}\int_0^L v^2 dx + \beta \frac{a^{2n-2}}{\overline{u}} \int_0^t \int_0^L v_x^2 dx ds$$

$$\leq C_1 + C_1(\varepsilon) \int_0^t \int_0^L \left(\theta^2 v^2 + \frac{\theta_x^2}{u\theta^2} + \theta v^2\right) dx ds + C_1\varepsilon \int_0^t \int_0^L \theta u_x^2 dx ds$$

$$\leq C_1 + C_1(\varepsilon) \int_0^t \sup_{x\in[0,L]} |v|^2 \int_0^L (\theta^2 + \theta) dx ds + C_1\varepsilon \int_0^t \int_0^L \theta u_x^2 dx ds$$

$$\leq C_1 + C_1\varepsilon \int_0^t \int_0^L \theta u_x^2 dx ds. \quad (1.2.49)$$

From (1.1.9) and (1.1.10), we derive

$$\beta\left(\frac{u_x}{u}\right)_t = \beta\frac{u_{tx}}{u} - \beta\frac{u_x u_t}{u^2} = r^{1-n}(v_t - f) + R\left(\frac{\theta}{u}\right)_x. \quad (1.2.50)$$

Multiplying (1.2.50) by $\frac{u_x}{u}$, then integrating the result over $[0,L] \times [0,t]$, we have

$$\beta\int_0^L \int_0^t \left(\frac{u_x}{u}\right)\left(\frac{u_x}{u}\right)_t ds dx$$

$$= \int_0^t \int_0^L r^{1-n}(v_t - f)\frac{u_x}{u} dx ds + R\int_0^t \int_0^L \frac{\theta_x}{u}\frac{u_x}{u} dx ds - R\int_0^t \int_0^L \frac{\theta u_x^2}{u^3} dx ds,$$

where

$$R \int_0^t \int_0^L \frac{\theta_x u_x}{u^2} dx ds = R \int_0^t \int_0^L \frac{\theta_x}{\sqrt{u\theta}} \frac{\sqrt{\theta} u_x}{u\sqrt{u}} dx ds \tag{1.2.51}$$

$$\leq \frac{R}{2} \int_0^t \int_0^L \frac{\theta u_x^2}{u^3} dx ds + \frac{R}{2} \int_0^t \int_0^L \frac{\theta_x^2}{u\theta} dx ds$$

$$\leq \frac{R}{2} \int_0^t \int_0^L \frac{\theta u_x^2}{u^3} dx ds + \frac{R}{4} \int_0^t \int_0^L \frac{\theta_x^2}{u} \left(1 + \frac{1}{\theta^2}\right) dx ds.$$

Using the mean value theorem and (1.2.51), we have

$$\frac{\beta}{2} \int_0^L \left(\frac{u_x}{u}\right)^2 dx + \frac{R}{2} \int_0^t \int_0^L \frac{\theta u_x^2}{u^3} dx ds$$

$$\leq C_1 + \int_0^t \int_0^L r^{1-n} v_t \frac{u_x}{u} dx ds + C_1 \int_0^t \int_0^L \frac{\theta_x^2}{u} \left(1 + \frac{1}{\theta^2}\right) dx ds$$

$$- \int_0^t \int_0^L r^{1-n} f \frac{u_x}{u} dx ds, \tag{1.2.52}$$

where

$$\left| \int_0^t \int_0^L r^{1-n} v_t \frac{u_x}{u} dx ds \right| \tag{1.2.53}$$

$$= \left| \int_0^L r^{1-n} v \frac{u_x}{u} \Big|_0^t dx + (n-1) \int_0^t \int_0^L r^{-n} v^2 \frac{u_x}{u} dx ds - \int_0^t \int_0^L r^{1-n} v \left(\frac{u_t}{u}\right)_x dx ds \right|$$

$$\leq C_1 + \frac{\beta}{4} \int_0^L \left(\frac{u_x}{u}\right)^2 dx + C_1 \int_0^t \sup_{x \in [0,L]} v^2 \left(\int_0^L \left(u_x^2 + \frac{1}{u^2}\right) dx\right) ds$$

$$+ \frac{2}{u} \int_0^t \int_0^L v_x^2 dx ds$$

$$\leq C_1 + \frac{\beta}{4} \int_0^L \left(\frac{u_x}{u}\right)^2 dx + C_1 \int_0^t \sup_{x \in [0,L]} v^2 \int_0^L u_x^2 dx ds + \frac{2}{u} \int_0^t \int_0^L v_x^2 dx ds,$$

$$\left| - \int_0^t \int_0^L r^{1-n} f \frac{u_x}{u} dx ds \right| \leq C_1 \int_0^t \int_0^L |f| u_x^2 dx ds + C_1 \int_0^t \int_0^L |f| dx ds$$

$$\leq C_1 + C_1 \int_0^t \sup_{x \in [0,L]} |f| \int_0^L u_x^2 dx ds. \tag{1.2.54}$$

Combining (1.2.51)–(1.2.54), and using Lemmas 1.2.2, 1.2.5 and 1.2.6, we get

$$\frac{\beta}{4(\overline{u})^2} \int_0^L u_x^2 dx + \frac{R}{2(\overline{u})^3} \int_0^t \int_0^L \theta u_x^2 dx ds \tag{1.2.55}$$

$$\leq C_1 + C_1 \int_0^t \left(\sup_{x \in [0,L]} v^2(\cdot, t) + \sup_{x \in [0,L]} |f| \right) \left(\int_0^L u_x^2 dx \right) ds + \frac{2}{u} \int_0^t \int_0^L v_x^2 dx ds.$$

Multiplying (1.2.55) by $\dfrac{u\beta a^{2n-2}}{4\bar{u}}$, then adding the result up to (1.2.49), by the smallness of ε, we derive

$$\int_0^L u_x^2 dx + \int_0^t \int_0^L (v_x^2 + \theta u_x^2)dxds \tag{1.2.56}$$

$$\leq C_1 + C_1 \int_0^t \left(\sup_{x\in[0,L]} v^2(x,t) + \sup_{x\in[0,L]} |f| \right) \int_0^L u_x^2 dxds, \quad \forall t > 0.$$

Applying Gronwall's inequality to (1.2.56), we get

$$\int_0^L u_x^2 dx + \int_0^t \int_0^L (v_x^2 + \theta u_x^2)dxds \leq C_1, \quad \forall t > 0. \tag{1.2.57}$$

Multiplying (1.2.39) by u_x^2, then integrating the result over $[0,t] \times [0,L]$, by (1.2.7), (1.2.32) and (1.2.57), we have

$$\frac{\alpha_1}{2} \int_0^t \int_0^L u_x^2 dxds \leq \int_0^t \int_0^L \theta u_x^2 dxds + C_1 \int_0^t \left(\int_0^L \frac{\theta_x^2}{u\theta^2}dx \right) \left(\int_0^L u_x^2 dx \right) ds \leq C_1. \tag{1.2.58}$$

Combining (1.2.57) and (1.2.58), we obtain (1.2.48). $\qquad\square$

Lemma 1.2.8. *Under the conditions of Theorem 1.2.1, we have*

$$\int_0^L v_x^2(x,t)dx + \int_0^t \int_0^L v_t^2(x,s)dxds \leq C_1, \quad \forall t > 0, \tag{1.2.59}$$

$$|v(x,t)| \leq C_1, \quad \forall(x,t) \in [0,L] \times [0,+\infty). \tag{1.2.60}$$

Proof. By the embedding theorem, we have

$$\sup_{x\in[0,L]} \theta(x,t) \leq C_1 + C_1\|\theta_x(t)\|, \quad \forall t > 0. \tag{1.2.61}$$

Multiplying (1.1.10) by v_t, integrating the result over $[0,L] \times [0,t]$, using the estimate

$$(r^{n-1}v_t)_x = (r^{n-1}v)_{xt} - (n-1)(r^{n-2}v^2)_x, \quad |(r^{n-2}v^2)_x| \leq C_1(v^2 + v_x^2),$$

we get

$$\int_0^t \int_0^L v_t^2 dxds = \int_0^t \int_0^L r^{n-1}v_t \left(\beta\frac{(r^{n-1}v)_x}{u} - R\frac{\theta}{u} \right)_x dxds + \int_0^t \int_0^L v_t f dxds, \tag{1.2.62}$$

$$\int_0^t \|v_t(s)\|^2 ds + \frac{\beta}{u}\|(r^{n-1}v)_x\|^2 \leq C_1 + (n-1)\int_0^t \int_0^L \frac{\beta}{u}(r^{n-1}v)_x(r^{n-2}v^2)_x dx ds$$

$$+ \int_0^t \int_0^L v_t f dx ds - R \int_0^t \int_0^L \left(\frac{\theta}{u}\right)_x (r^{n-1}v_t) dx ds$$

$$\leq C_1 + C_1 \int_0^t \int_0^L (|v| + |v_x|)(v^2 + v_x^2) dx ds + C_1 \int_0^t \int_0^L f^2 dx ds$$

$$+ \frac{1}{2}\int_0^t \|v_t(s)\|^2 ds + C_1 \int_0^t \int_0^L (\theta_x^2 + \theta^2 u_x^2) dx ds \tag{1.2.63}$$

where

$$\int_0^t \int_0^L \theta^2 u_x^2 dx ds \leq \int_0^t \sup_{x \in [0,L]} \theta^2 \left(\int_0^L u_x^2 dx\right) ds$$

$$\leq \int_0^t (C_1 + C_1\|\theta_x\|^2)\left(\int_0^L u_x^2 dx\right) ds$$

$$\leq C_1 \int_0^t \int_0^L u_x^2 dx ds + C_1 \int_0^t \int_0^L \theta_x^2 dx ds \leq C_1,$$

$$I \triangleq \int_0^t \int_0^L (|v| + |v_x|)(v^2 + v_x^2) dx ds$$

$$\leq C_1 \int_0^t (\|v\|_{L^\infty} + \|v_x\|_{L^\infty})(\|v\|^2 + \|v_x\|^2) ds$$

$$\leq C_1 \int_0^t \left[\|v\|_{L^\infty}\|v\|^2 + \|v\|_{L^\infty}\|v_x\|^2 + \|v_x\|_{L^\infty}\|v\|^2 + \|v_x\|_{L^\infty}\|v_x\|^2\right](s) ds$$

$$\leq C_1 \int_0^t \left(\|v\|_{L^\infty}^2 + \|v_x\|_{L^\infty}^2\right)(s) ds$$

$$+ C_1 \left(\int_0^t \|v(s)\|_{L^\infty}^2 ds\right)^{\frac{1}{2}}\left(\int_0^t \|v_x(s)\|^2 ds\right)^{\frac{1}{2}} \cdot \sup_{s \in [0,t]} \|v_x(s)\|$$

$$+ C_1 \left(\int_0^t \|v_x(s)\|_{L^\infty}^2 ds\right)^{\frac{1}{2}}\left(\int_0^t \|v_x(s)\|^2 ds\right)^{\frac{1}{2}} \cdot \sup_{s \in [0,t]} \|v_x(s)\|$$

$$\leq C_1 + C_1 \sup_{s \in [0,t]} \|v_x(s)\| + C_1 \left(\int_0^t \|v_{xx}(s)\|^2 ds\right)^{\frac{1}{2}} \cdot \sup_{s \in [0,t]} \|v_x(s)\|. \tag{1.2.64}$$

By (1.1.10), we have

$$\|v_{xx}\| \leq C_1\Big[\|v_t\| + \|(r^{n-1})_{xx}v\| + \|(r^{n-1})_x v_x\| + \|\theta_x\| + \|u_x\| + \|u_x uv\|$$

$$+ \|u_x v_x\| + \|\theta u_x\| + \|f\|\Big]$$

$$\leq C_1\Big(\|v_t\| + \|v\|_{L^\infty}\|u_x\| + \|v_x\| + \|\theta_x\| + \|u_x\| + \|v_x\|_{L^\infty}\|u_x\|$$
$$+ \|\theta u_x\| + \|f\|\Big)$$
$$\leq \frac{1}{2}\|v_{xx}\| + C_1\Big(\|v_t\| + \|v\|_{L^\infty} + \|v_x\| + \|\theta_x\| + \|u_x\| + \|\theta u_x\| + \|f\|\Big),$$

i.e.,

$$\|v_{xx}\| \leq C_1(\|v_t\| + \|v\|_{L^\infty} + \|v_x\| + \|\theta_x\| + \|u_x\| + \|\theta u_x\| + \|f\|). \tag{1.2.65}$$

Combining (1.2.64) and (1.2.65), using Young's inequality, we have

$$I \leq C_1 \sup_{s\in[0,t]} \|v_x(s)\| + C_1 \left\{ 1 + \left(\int_0^t \|v_t(s)\|^2 ds\right)^{\frac{1}{2}} \right\} \sup_{s\in[0,t]} \|v_x(s)\|$$

$$\leq \varepsilon \sup_{s\in[0,t]} \|v_x(s)\|^2 + \varepsilon \int_0^t \|v_t(s)\|^2 ds + C_1\varepsilon^{-1}, \quad \forall \varepsilon > 0. \tag{1.2.66}$$

Combining (1.2.64) and (1.2.66), picking ε small enough, we derive (1.2.59).

By the embedding theorem, (1.2.7) and (1.2.59), we have

$$|v(x,t)| \leq C_1\|v\|_{L^\infty[0,L]} \leq C_1\|v\|_{H^1[0,L]} \leq C_1. \qquad \square$$

Lemma 1.2.9. *Under the conditions of Theorem 1.2.1, we have*

$$\int_0^t \|v_x(s)\|_{L^\infty}^2 ds + \int_0^t \int_0^L (u_{xt}^2 + v_{xx}^2)(x,s)dxds \leq C_1, \quad \forall t > 0. \tag{1.2.67}$$

Proof. By (1.1.9) and (1.1.10), we deduce after a direct calculation that

$$\int_0^t \int_0^L (u_{xt}^2 + v_{xx}^2)dxds$$

$$\leq C_1 \int_0^t \int_0^L \left((r^{n-1}v)_{xx}^2 + v_t^2 + v_x^2u_x^2 + \theta_x^2 + \theta^2u_x^2 + f^2\right) dxds$$

$$\leq C_1 + C_1 \int_0^t \int_0^L \left[\frac{(r^{n-1}v)_x}{u}\right]_x^2 dxds + C_1 \int_0^t \sup_{x\in[0,L]} \theta^2 \left(\int_0^L u_x^2 dx\right) ds$$

$$\leq C_1 + C_1 \int_0^t \int_0^L (v_t^2 + \theta_x^2 + \theta^2u_x^2 + f^2)dxds$$

$$\leq C_1.$$

By the embedding theorem, we have

$$\int_0^t \|v_x(s)\|_{L^\infty}^2 ds \leq C \int_0^t \int_0^L v_{xx}^2(x,s)dxds \leq C_1.$$

By (1.2.7), (1.2.41), (1.2.48), (1.2.59) and (1.1.16), we obtain (1.2.67). $\qquad \square$

Lemma 1.2.10. *Under the conditions of Theorem 1.2.1, we have*

$$\int_0^L \theta_x^2(x,t)dx + \int_0^t \int_0^L (\theta_t^2 + \theta_{xx}^2)(x,s)dxds \leq C_1, \quad \forall t > 0, \qquad (1.2.68)$$

$$\theta(x,t) \leq C_1, \quad \forall(x,t) \in [0,L] \times [0,+\infty). \qquad (1.2.69)$$

Proof. Multiplying (1.1.11) by θ_t, then integrating the resultant over $[0,L] \times [0,t]$, we have

$$C_v \int_0^t \int_0^L \theta_t^2 dxds = k \int_0^t \int_0^L \left(\frac{r^{2n-2}\theta_x}{u}\right)_x \theta_t dxds + \int_0^t \int_0^L \frac{\beta}{u}(r^{n-1}v)_x^2 \theta_t dxds$$

$$+ \int_0^t \int_0^L \theta_t g dxds - 2\mu(n-1)\int_0^t \int_0^L (r^{n-2}v^2)_x \theta_t dxds$$

$$- \int_0^t \int_0^L R\frac{\theta}{u}(r^{n-1}v)_x \theta_t dxds \qquad (1.2.70)$$

where

$$\int_0^t \int_0^L \frac{\beta}{u}(r^{n-1}v)_x^2 \theta_t dxds \qquad (1.2.71)$$

$$\leq C_1 \int_0^t \int_0^L \theta_t^2 dxds + C_1 \int_0^t \int_0^L |(r^{n-1}v)_x|^2(v^2 + v_x^2)dxds$$

$$\leq C_1 \int_0^t \int_0^L \theta_t^2 dxds + C_1 \int_0^t \left(\sup_{x\in[0,L]}|v|^2 + \sup_{x\in[0,L]}|v_x|^2\right)\int_0^L |(r^{n-1}v)_x|^2 dxds,$$

$$k \int_0^t \int_0^L \left(\frac{r^{2n-2}\theta_x}{u}\right)_x \theta_t dxds \qquad (1.2.72)$$

$$= k \int_0^t \left(\frac{r^{2n-2}\theta_x}{u}\right)\theta_t \Big|_0^L ds - k \int_0^t \int_0^L \left(\frac{r^{2n-2}\theta_x}{u}\right)\theta_{tx} dxds$$

$$= -k \int_0^L \frac{r^{2n-2}\theta_x}{u}\theta_x \Big|_0^t dx + k \int_0^t \int_0^L \left(\frac{r^{2n-2}\theta_x}{u}\right)_t \theta_x dxds$$

$$\leq C_1 - k \int_0^L \frac{r^{2n-2}}{u}\theta_x^2 dx$$

$$+ k \int_0^t \int_0^L \left\{\frac{(2n-2)r^{2n-3}v\theta_x^2}{u} + \frac{r^{2n-2}\theta_{tx}\theta_x}{u} - \frac{r^{2n-2}\theta_x^2 u_t}{u^2}\right\}(x,s)dxds,$$

$$- 2\mu(n-1)\int_0^t \int_0^L (r^{n-2}v^2)_x \theta_t dxds \qquad (1.2.73)$$

$$= -2\mu(n-1)\int_0^t (r^{n-2}v^2)\theta_t \Big|_0^L ds + 2\mu(n-1)\int_0^t \int_0^L (r^{n-2}v^2)\theta_{tx} dxds$$

$$= 2\mu(n-1)\int_0^L (r^{n-2}v^2)\theta_x \Big|_0^t dx$$

$$- 2\mu(n-1)\int_0^t \int_0^L [(n-2)r^{n-3}v^3 + 2r^{n-2}vv_t]\theta_x dxds,$$

$$\left| -\int_0^t \int_0^L R\frac{\theta}{u}(r^{n-1}v)_x\theta_t dxds \right| \tag{1.2.74}$$

$$\leq C_1 \int_0^t \int_0^L \theta^2 v_x^2 dxds + \frac{C_v}{4}\int_0^t \int_0^L \theta_t^2 dxds$$

$$\leq C_1 \int_0^t \int_0^L (\theta - \bar\theta)^2 v_x^2 dxds + C_1 \int_0^t \int_0^L v_x^2 dxds + \frac{C_v}{4}\int_0^t \int_0^L \theta_t^2 dxds$$

$$\leq C_1 + C_1 \int_0^t \sup_{x\in[0,L]}(\theta - \bar\theta)^2 \int_0^L v_x^2 dxds + \frac{C_v}{4}\int_0^t \int_0^L \theta_t^2 dxds$$

$$\leq C_1 + C_1 \int_0^t \|\theta_x(s)\|^2 ds + \frac{C_v}{4}\int_0^t \int_0^L \theta_t^2 dxds$$

$$\leq C_1 + \frac{C_v}{4}\int_0^t \int_0^L \theta_t^2 dxds,$$

$$\left| \int_0^t \int_0^L \theta_t g dxds \right| \leq C_1 \int_0^t \int_0^L g^2 dxds + \frac{C_v}{4}\int_0^t \int_0^L \theta_t^2 dxds. \tag{1.2.75}$$

Combining (1.2.70)–(1.2.75), and using Young's inequality, (1.2.32), (1.2.36), (1.2.59) and (1.1.17), we can obtain that (1.2.68). (1.2.69) follows from (1.2.68) and (1.2.61). $\qquad\square$

Proof of Theorem 1.2.1. Using Lemmas 1.2.1–1.2.10, the embedding theorem and noting that

$$r_x = r^{1-n}u, \quad r_{xx} = (1-n)r^{-n}r_x u + r^{1-n}u_x,$$

$$r_t = v, \quad r_{tx} = v_x, \quad r_{txx} = v_{xx},$$

we can obtain (1.2.3). $\qquad\square$

1.3 Global Existence of Solutions in H^2

For external force $f(r,t)$ and heat source $g(r,t)$, besides (1.1.16)–(1.1.17), we suppose that $f(r,t)$, $g(r,t)$ are C^1 functions on $[a,b]\times\mathbb{R}^+$ satisfying

$$f(r,t)\in L^\infty(\mathbb{R}^+, L^2[a,b]),\ f_r\in L^2(\mathbb{R}^+, L^2[a,b]),\ f_t\in L^2(\mathbb{R}^+, L^2[a,b]), \tag{1.3.1}$$

$$g(r,t)\in L^\infty(\mathbb{R}^+, L^2[a,b]),\ g_r\in L^2(\mathbb{R}^+, L^2[a,b]),\ g_t\in L^2(\mathbb{R}^+, L^2[a,b]). \tag{1.3.2}$$

Constant C_2 denotes a generic positive constant depending only on the H^2 norm of initial data (u_0, v_0, θ_0),

$$\|f\|_{L^\infty(\mathbb{R}^+, L^2[a,b])}, \quad \|f_r\|_{L^2(\mathbb{R}^+, L^2[a,b])}, \quad \|f_t\|_{L^2(\mathbb{R}^+, L^2[a,b])},$$
$$\|g\|_{L^\infty(\mathbb{R}^+, L^2[a,b])}, \quad \|g_r\|_{L^2(\mathbb{R}^+, L^2[a,b])}, \quad \|g_t\|_{L^2(\mathbb{R}^+, L^2[a,b])}$$

and constant C_1, but is independent of t.

Remark 1.3.1. By Remark 1.2.1 and (1.3.1)–(1.3.2), we easily know that assumptions (1.3.1)–(1.3.2) are equivalent to the following conditions: $\forall x \in [0, L]$,

$$f(r(x,t),t) \in L^\infty(\mathbb{R}^+, L^2[0, L]),$$
$$f_r(r(x,t),t) \in L^2(\mathbb{R}^+, L^2[0, L]), \quad f_t(r(x,t),t) \in L^2(\mathbb{R}^+, L^2[0, L]), \qquad (1.3.3)$$
$$g(r(x,t),t) \in L^\infty(\mathbb{R}^+, L^2[a, b]), \quad g_r(r(x,t)) \in L^2(\mathbb{R}^+, L^2[0, L]),$$
$$g_t(r(x,t),t) \in L^2(\mathbb{R}^+, L^2[0, L]). \qquad (1.3.4)$$

Therefore the generic positive constant C_2 depends only on the norm of initial data (u_0, v_0, θ_0) in H^2 and the norms of f and g in the class of functions in (1.3.3)–(1.3.4), but is independent of t.

Theorem 1.3.1. *Under the conditions* (1.1.16), (1.1.17) (1.3.1), (1.3.2), *if*

$$(u_0, v_0, \theta_0) \in H^2[0, L] \times H_0^2[0, L] \times H^2[0, L],$$

problem (1.1.9)–(1.1.15) *admits a unique global solution*

$$(u(t), v(t), \theta(t)) \in C([0, +\infty), H^2[0, L] \times H_0^2[0, L] \times H^2[0, L])$$

satisfying that for any $t > 0$,

$$\|r(t) - \bar{r}\|_{H^3}^2 + \|r_t(t)\|_{H^2}^2 + \|u(t) - \bar{u}\|_{H^2}^2 + \|v(t) - \bar{v}\|_{H^2}^2 + \|\theta(t) - \bar{\theta}\|_{H^2}^2$$
$$+ \|u(t) - \bar{u}\|_{W^{1,\infty}}^2 + \|v(t) - \bar{v}\|_{W^{1,\infty}}^2 + \|\theta(t) - \bar{\theta}\|_{W^{1,\infty}}^2 + \|u_t(t)\|_{H^1}^2$$
$$+ \|v_t(t)\|^2 + \|\theta_t(t)\|^2 + \int_0^t \Big(\|u - \bar{u}\|_{H^2}^2 + \|u_t\|_{H^2}^2 + \|v - \bar{v}\|_{H^3}^2 $$
$$+ \|\theta - \bar{\theta}\|_{H^3}^2 + \|v_t\|_{H^1}^2 + \|\theta_t\|_{H^1}^2 + \|r - \bar{r}\|_{H^3}^2 + \|r_t\|_{H^3}^2 \Big)(s)ds \le C_2. \quad (1.3.5)$$

The proof of Theorem 1.3.1 can be divided into the following several lemmas.

Lemma 1.3.1. *Under the conditions of Theorem 1.3.1, we have that for any* $t > 0$,

$$\|v_t(t)\|^2 + \|v_{xx}(t)\|^2 + \|v_x(t)\|_{L^\infty}^2 + \int_0^t \|v_{tx}(s)\|^2 ds \le C_2, \qquad (1.3.6)$$

$$\|\theta_t(t)\|^2 + \|\theta_{xx}(t)\|^2 + \|\theta_x(t)\|_{L^\infty}^2 + \int_0^t \|\theta_{tx}(s)\|^2 ds \le C_2. \qquad (1.3.7)$$

Proof. Differentiating (1.1.10) with respect to t, we have

$$v_{tt} = (n-1)r^{n-2}r_t \left(\beta \frac{(r^{n-1}v)_x}{u} - R\frac{\theta}{u} \right)_x + r^{n-1} \left(\beta \frac{(r^{n-1}v)_x}{u} - R\frac{\theta}{u} \right)_{tx} + \frac{df}{dt}.$$
$$(1.3.8)$$

Multiplying (1.3.8) by v_t, then integrating the result over $[0,t] \times [0,L]$, by (1.1.9) and Theorem 1.2.1, and noting that $\frac{df}{dt} = f_r v + f_t$, we can derive

$$\int_0^t \int_0^L v_{tt} v_t \, dx \, ds \geq \frac{1}{2} \int_0^L v_t^2 \, dx - C_2,$$
$$(1.3.9)$$

$$\int_0^t \int_0^L v_t r^{n-1} \left(\beta \frac{(r^{n-1}v)_x}{u} - R\frac{\theta}{u} \right)_{tx} dx \, ds$$

$$= \int_0^t \int_0^L v_t r^{n-1} \left(\beta \frac{(r^{n-1}v)_{tx}}{u} - \frac{(r^{n-1}v)_x u_t}{u^2} - R\frac{\theta_t}{u} + R\frac{\theta u_t}{u^2} \right)_x dx \, ds$$

$$\leq -C_2^{-1} \int_0^t \int_0^L v_{tx}^2 \, dx \, ds + C_2(\varepsilon) \int_0^t \int_0^L v_{tx}^2 \, dx \, ds + C_2 \int_0^t \int_0^L (r^{n-1}v)_x^4 \, dx \, ds$$

$$+ \int_0^t \int_0^L \left[\theta_t^2 + \theta^2 (r^{n-1}v)_x^2 \right] dx \, ds$$

$$\leq C_2 - C_2^{-1} \int_0^t \int_0^L v_{tx}^2 \, dx \, ds + C_2 \int_0^t \sup_{x\in[0,L]} v_x^2 \left(\int_0^L v_x^2 \, dx \right) ds$$

$$+ C_2 \int_0^t \sup_{x\in[0,L]} \theta^2 \left(\int_0^L v_x^2 \, dx \right) ds$$

$$\leq C_2 - C_2^{-1} \int_0^t \int_0^L v_{tx}^2 \, dx \, ds,$$
$$(1.3.10)$$

$$\int_0^t \int_0^L (n-1)r^{n-2} v v_t \left(\beta \frac{(r^{n-1}v)_x}{u} - R\frac{\theta}{u} \right)_x dx \, ds$$

$$\leq C_2 \int_0^t \int_0^L \left(v_t^2 + v_x^2 u_x^2 + v_{xx}^2 + \theta_x^2 u_x^2 + \theta_x^2 \right) dx \, ds$$

$$\leq C_2 + C_2 \int_0^t \sup_{x\in[0,L]} v_x^2 \left(\int_0^L u_x^2 \, dx \right) ds + C_2 \int_0^t \sup_{x\in[0,L]} \theta_x^2 \left(\int_0^L u_x^2 \, dx \right) ds$$

$$\leq C_2 + C_2 \int_0^t \left(\|v_x\|^2 + \|v_{xx}\|^2 \right)(s) ds + C_2 \int_0^t \left(\|\theta_x\|^2 + \|\theta_{xx}\|^2 \right)(s) ds$$

$$\leq C_2,$$
$$(1.3.11)$$

$$\left| \int_0^t \int_0^L \frac{df}{dt} v_t \, dx \, ds \right| \leq C_1 \int_0^t \int_0^L v_t^2 \, dx \, ds + C_1 \int_0^t \int_0^L (f_r^2 + f_t^2) \, dx \, ds \leq C_2.$$
$$(1.3.12)$$

Combining (1.3.8)–(1.3.12), we obtain

$$\|v_t(t)\|^2 + \int_0^t \|v_{tx}(s)\|^2 ds \leq C_2 \tag{1.3.13}$$

which, by noting that from (1.1.10)

$$\|v_{xx}(t)\| \leq C_2(\|v_t(t)\| + \|u_x(t)\| + \|v_x(t)\|_{H^1} + \|\theta(t)\|_{H^1} + \|f\|) \leq C_2,$$

gives

$$\|v_{xx}(t)\| \leq C_2. \tag{1.3.14}$$

Hence (1.3.6) follows from (1.3.13) by using the embedding theorem.

Similarly, differentiating (1.1.11) with respect to t, we can get

$$C_v \theta_{tt} = k \left(\frac{r^{2n-2}\theta_x}{u} \right)_{tx} + \left(\beta \frac{(r^{n-1}v)_x}{u} - R\frac{\theta}{u} \right)_t (r^{n-1}v)_x$$

$$+ \left(\beta \frac{(r^{n-1}v)_x}{u} - R\frac{\theta}{u} \right)(r^{n-1}v)_{tx} - 2\mu(n-1)(r^{n-2}v^2)_{tx} + \frac{dg}{dt}. \tag{1.3.15}$$

Multiplying (1.3.15) by θ_t, then integrating the result over $[0,t] \times [0,L]$, and noting that

$$\frac{dg}{dt} = g_r v + g_t,$$

we have

$$\int_0^t \int_0^L C_v \theta_t \theta_{tt} dx ds \geq \frac{C_v}{2} \int_0^L \theta_t^2 dx - C_2, \tag{1.3.16}$$

$$\int_0^t \int_0^L k\theta_t \left(\frac{r^{2n-2}\theta_x}{u} \right)_{tx} dx ds$$

$$\leq -C_2^{-1} \int_0^t \int_0^L \theta_{tx}^2 dx ds + C_2(\varepsilon) \int_0^t \int_0^L \theta_{tx}^2 dx ds + C_2 \int_0^t \int_0^L \theta_x^2 u_t^2 dx ds$$

$$\leq -C_2^{-1} \int_0^t \int_0^L \theta_{tx}^2 dx ds + C_2 \int_0^t \sup_{x \in [0,L]} v_x^2 \int_0^L \theta_x^2 dx ds$$

$$\leq C_2 - C_2^{-1} \int_0^t \int_0^L \theta_{tx}^2 dx ds, \tag{1.3.17}$$

$$\int_0^t \int_0^L \left(\beta \frac{(r^{n-1}v)_x}{u} - R\frac{\theta}{u} \right)_t (r^{n-1}v)_x \theta_t dx ds$$

$$\leq C_2 \int_0^t \int_0^L (v_{tx}^2 + v_x^4 + \theta_t^2 + \theta^2 v_x^2 + v_x^2 \theta_t^2) dx ds$$

$$\leq C_2 + C_2 \int_0^t \sup_{x \in [0,L]} (v_x^2 + \theta^2) \left(\int_0^L v_x^2 dx \right) ds + \int_0^t \sup_{x \in [0,L]} v_x^2 \left(\int_0^L \theta_t^2 dx \right) ds$$

$$\leq C_2 + C_2 \int_0^t \sup_{x\in[0,L]} v_x^2 \left(\int_0^L \theta_t^2 dx \right) ds. \tag{1.3.18}$$

From (1.1.14) and Lemmas 1.2.1, 1.2.5 and 1.2.7–1.2.10, we derive

$$\left\| \beta \frac{(r^{n-1}v)_x}{u} - R\frac{\theta}{u} \right\|_{L^\infty} \leq C_2, \tag{1.3.19}$$

$$\|(r^{n-1}v)_{tx}\| \leq C_2(\|v_x\| + \|v_{tx}\|), \tag{1.3.20}$$

$$\left| \int_0^L \theta_t dx \right| \leq C_1 \left| \int_0^L \frac{1}{u} \left(\beta(r^{n-1}x)_x - R\theta \right)(r^{n-1}v)_x \right|$$
$$\leq C_2\|v_x\| \tag{1.3.21}$$

which, with the Poincaré inequality, gives

$$\|\theta_t\| \leq \left| \int_0^L \theta_t dx \right| + C\|\theta_{tx}\| \leq C_2(\|v_x\| + \|\theta_{tx}\|). \tag{1.3.22}$$

Thus from (1.3.19)–(1.3.22) we derive for $\varepsilon > 0$ small enough,

$$\int_0^t \int_0^L \left(\beta\frac{(r^{n-1}v)_x}{u} - R\frac{\theta}{u} \right)(r^{n-1}v)_{tx}\theta_t dx ds$$
$$\leq C_2\|(r^{n-1}v)_{tx}\|\|\theta_t\|$$
$$\leq C_2(\|v_x\| + \|v_{tx}\|)(\|v_x\| + \|\theta_{tx}\|)$$
$$\leq \varepsilon\|\theta_{tx}\|^2 + C_2(\|v_x\|^2 + \|v_{tx}\|^2), \tag{1.3.23}$$

$$\int_0^t \int_0^L -2\mu(n-1)(r^{n-2}v^2)_{tx}\theta_t dx ds$$
$$\leq C_2 \int_0^t \int_0^L (\theta_t^2 + v_x^2v_t^2 + v^2v_{tx}^2)dx ds$$
$$\leq C_2 + C_2 \int_0^t \sup_{x\in[0,L]} v_t^2 \left(\int_0^L v_x^2 dx \right) ds + C_2 \int_0^t \int_0^L v_{tx}^2 dx ds$$
$$\leq C_2, \tag{1.3.24}$$

$$\int_0^t \int_0^L \frac{dg}{dt}\theta_t dx ds \leq C_1 \int_0^t \int_0^L (g_r^2 + g_t^2 + \theta_t^2)dx ds \leq C_2. \tag{1.3.25}$$

Combining (1.3.15)–(1.3.25), we derive for $\varepsilon > 0$ small enough,

$$\int_0^L \theta_t^2 dx + \int_0^t \int_0^L \theta_{tx}^2 dx ds \leq C_2 + C_2 \int_0^t \sup_{x\in[0,L]} v_x^2 \left(\int_0^L \theta_t^2 dx \right) ds.$$

By Gronwall's inequality, we obtain

$$\|\theta_t(t)\|^2 + \int_0^t \|\theta_{tx}(s)\|^2 ds \le C_2, \tag{1.3.26}$$

which, by noting that from (1.1.11)

$$\|\theta_{xx}(t)\| \le C_2(\|\theta_t(t)\| + \|\theta(t)\|_{H^1} + \|u(t)\|_{H^1} + \|v(t)\|_{H^1} + \|g(t)\|) \le C_2,$$

and by the embedding theorem, implies (1.3.7). □

Lemma 1.3.2. *Under the conditions of Theorem 1.3.1, we have*

$$\|u_{xx}(t)\|^2 + \int_0^t \|u_{xx}(s)\|^2 ds \le C_2, \quad \forall t > 0. \tag{1.3.27}$$

Proof. Differentiating (1.1.10) with respect to x, we have

$$\beta \frac{d}{dt}\left(\frac{u_{xx}}{u}\right) + \frac{R\theta u_{xx}}{u^2} = r^{1-n}v_{tx} + (1-n)r^{1-2n}uv_t + 2\beta\frac{(r^{n-1}v)_{xx}u_x}{u^2}$$
$$- 2\beta\frac{(r^{n-1}v)_x u_x^2}{u^3} + R\frac{\theta_{xx}}{u} - 2R\frac{\theta_x u_x}{u^2} + 2R\frac{\theta u_x^2}{u^3} - r^{1-n}f_r u - (1-n)r^{1-2n}uf$$
$$:= M \tag{1.3.28}$$

where

$$\|M\| \le C_2(\|v_{tx}\| + \|v_t\| + \|u_x\| + \|v_x\|_{H^1} + \|\theta\|_{H^2} + \|f_r\| + \|f\|).$$

By Theorem 1.2.1, condition (1.3.1) and Lemma 1.3.1, we get

$$\int_0^t \|M(s)\|^2 ds \le C_2, \quad \forall t > 0. \tag{1.3.29}$$

Multiplying (1.3.28) by $\frac{u_{xx}}{u}$, then integrating the result over $[0,t] \times [0,L]$, using Young's inequality and (1.3.29), we can obtain (1.3.27). □

Lemma 1.3.3. *Under conditions of Theorem 1.3.1, we have*

$$\int_0^t \left(\|v_{xxx}\|^2 + \|\theta_{xxx}\|^2\right)(s)ds \le C_2, \quad \forall t > 0. \tag{1.3.30}$$

Proof. Differentiating (1.1.10) with respect to x, we can get

$$\|v_{xxx}(t)\| \le C_2\left(\|v_{tx}\| + \|v_x\|_{H^1} + \|u_x\|_{H^1} + \|\theta\|_{H^2} + \|f_r u\|\right).$$

By Theorem 1.2.1, Lemmas 1.3.1–1.3.2, we have

$$\int_0^t \|v_{xxx}(s)\|^2 ds \le C_2. \tag{1.3.31}$$

Similarly, differentiating (1.1.11) with respect to x, we can obtain

$$
C_v \theta_{tx} - k\frac{(r^{2n-2}\theta_x)_{xx}}{u} + 2k\frac{(r^{2n-2}\theta_x)_x u_x}{u^2} + k\frac{(r^{2n-2}\theta_x)u_{xx}}{u^2} - 2k\frac{(r^{2n-2}\theta_x)u_x^2}{u^3}
$$
$$
- \beta\frac{[(r^{n-1}v)_x]_x^2}{u} + \beta\frac{(r^{n-1}v)_x^2 u_x}{u^2} + R\frac{\theta_x(r^{n-1}v)_x + \theta(r^{n-1}v)_{xx}}{u}
$$
$$
- R\frac{\theta(r^{n-1}v)_x u_x}{u^2} + 2\mu(n-1)(r^{n-2}v^2)_{xx}
$$
$$
= g_r u r^{1-n},
$$

$$
\|\theta_{xxx}(t)\| \le C_2 \left(\|\theta_{tx}\| + \|\theta_x\|_{H^1} + \|u_x\|_{H^1} + \|v_x\|_{H^1} + \|g_r u\| \right).
$$

By Theorem 1.2.1, condition (1.3.1), (1.3.2) and Lemmas 1.3.1, 1.3.2, we derive (1.3.30). □

Proof of Theorem 1.3.1. By equations (1.1.9)–(1.1.11), Theorem 1.2.1, Lemmas 1.3.1–1.3.3 and the embedding theorem, we have

$$
\|u_{tx}(t)\| \le C_2\|v(t)\|_{H^2} \le C_2, \quad \|u_{txx}(t)\| \le C_2\|v(t)\|_{H^3},
$$
$$
\int_0^t \|u_{txx}(s)\|^2 ds \le C_2 \int_0^t \|v(s)\|_{H^3}^2 ds \le C_2,
$$
$$
\|r(t)\|_{H^3} \le C_2\|u(t)\|_{H^2} \le C_2, \quad \|r_t(t)\|_{H^2} \le C_2\|v(t)\|_{H^2} \le C_2,
$$
$$
\|u(t)\|_{W^{1,\infty}} \le C_2\|u(t)\|_{H^2} \le C_2, \quad \|v(t)\|_{W^{1,\infty}} \le C_2\|v(t)\|_{H^2} \le C_2,
$$
$$
\|\theta(t)\|_{W^{1,\infty}} \le C_2\|\theta(t)\|_{H^2} \le C_2.
$$

Combining Lemmas 1.3.1–1.3.3 with the above estimates, we obtain (1.3.5). □

1.4 Global Existence of Solutions in H^4

For external force $f(r,t)$ and heat source $g(r,t)$, besides (1.1.16)–(1.1.17) and (1.3.1)–(1.3.2), we suppose that $f(r,t)$, $g(r,t)$ are C^3 functions on $[a,b] \times \mathbb{R}^+$ satisfying

$$
f_{rr}, f_{rt}, f_{tt}, f_{rrr} \in L^2(\mathbb{R}^+, L^2[a,b]), \quad f_r, f_t, f_{rr} \in L^\infty(\mathbb{R}^+, L^2[a,b]), \quad (1.4.1)
$$
$$
g_{rr}, g_{rt}, g_{tt}, g_{rrr} \in L^2(\mathbb{R}^+, L^2[a,b]), \quad g_r, g_t, g_{rr} \in L^\infty(\mathbb{R}^+, L^2[a,b]). \quad (1.4.2)
$$

Constant C_4 denotes a generic positive constant depending only on the H^4 norm of initial data (u_0, v_0, θ_0),

$$
\|f_{rr}\|_{L^2(\mathbb{R}^+, L^2[a,b])}, \|f_{rt}\|_{L^2(\mathbb{R}^+, L^2[a,b])}, \|f_{tt}\|_{L^2(\mathbb{R}^+, L^2[a,b])}, \|f_{rrr}\|_{L^2(\mathbb{R}^+, L^2[a,b])},
$$
$$
\|g_{rr}\|_{L^2(\mathbb{R}^+, L^2[a,b])}, \|g_{rt}\|_{L^2(\mathbb{R}^+, L^2[a,b])}, \|g_{tt}\|_{L^2(\mathbb{R}^+, L^2[a,b])}, \|g_{rrr}\|_{L^2(\mathbb{R}^+, L^2[a,b])},
$$

and constants C_1, C_2, but is independent of t.

Remark 1.4.1. By Remarks 1.2.1 and 1.3.1, we easily know that assumptions (1.4.1)–(1.4.2) are equivalent to the following conditions: $\forall x \in [0, L]$,

$$f_{rr}(r(x,t),t), \ f_{rt}(r(x,t),t), \ f_{tt}(r(x,t),t),$$

$$f_{rrr}(r(x,t),t) \in L^2(\mathbb{R}^+, L^2[0,L]), \qquad (1.4.3)$$

$$f_r(r(x,t),t), \ f_t(r(x,t),t), \ f_{rr}(r(x,t),t) \in L^\infty(\mathbb{R}^+, L^2[0,L]), \qquad (1.4.4)$$

$$g_{rr}(r(x,t),t), \ g_{rt}(r(x,t),t), \ g_{tt}(r(x,t),t),$$

$$g_{rrr}(r(x,t),t) \in L^2(\mathbb{R}^+, L^2[0,L]), \qquad (1.4.5)$$

$$g_r(r(x,t),t), \ g_t(r(x,t),t), \ g_{rr}(r(x,t),t) \in L^\infty(\mathbb{R}^+, L^2[0,L]). \qquad (1.4.6)$$

Theorem 1.4.1. *Under conditions* (1.1.16), (1.1.17), (1.3.1), (1.3.2), (1.4.1) *and* (1.4.2), *if* $(u_0, v_0, \theta_0) \in H^4[0,L] \times H_0^4[0,L] \times H^4[0,L]$, *problem* (1.1.9)–(1.1.15) *admits a unique global solution* $(u(t), v(t), \theta(t)) \in C([0,+\infty), H^4[0,L] \times H_0^4[0,L] \times H^4[0,L])$ *satisfying, for any* $t > 0$,

$$\|u(t) - \bar{u}\|_{H^4}^2 + \|u_t(t)\|_{H^3}^2 + \|u_{tt}(t)\|_{H^1}^2 + \|u_{xx}(t)\|_{W^{1,\infty}}^2 + \|v(t)\|_{H^4}^2$$
$$+ \|v_t(t)\|_{H^2}^2 + \|v_{tt}(t)\|^2 + \|v_{xx}(t)\|_{W^{1,\infty}}^2 + \|\theta(t) - \bar{\theta}\|_{H^4}^2 + \|\theta_t(t)\|_{H^2}^2$$
$$+ \|\theta_{tt}(t)\|^2 + \|\theta_{xx}(t)\|_{W^{1,\infty}}^2 \le C_4, \qquad (1.4.7)$$

$$\int_0^t \Big\{ \|u - \bar{u}\|_{H^4}^2 + \|u_t\|_{H^4}^2 + \|u_{tt}\|_{H^2}^2 + \|u_{ttt}\|^2 + \|u_{xx}\|_{W^{1,\infty}}^2 + \|v\|_{H^5}^2$$
$$+ \|v_t\|_{H^3}^2 + \|v_{tt}\|_{H^1}^2 + \|v_{xx}\|_{W^{2,\infty}}^2 + \|\theta - \bar{\theta}\|_{H^5}^2 + \|\theta_t\|_{H^3}^2 + \|\theta_{tt}\|_{H^1}^2$$
$$+ \|\theta_{xx}\|_{W^{2,\infty}}^2 \Big\}(\tau)d\tau \le C_4. \qquad (1.4.8)$$

The proof of Theorem 1.4.1 can be done by a series of lemmas as follows.

Lemma 1.4.1. *Under conditions of Theorem 1.4.1, it holds that*

$$\|v_{tt}(t)\|^2 + \int_0^t \|v_{ttx}(\tau)\|^2 d\tau$$

$$\le C_4 + C_4 \int_0^t (\|v_{txx}\|^2 + \varepsilon\|\theta_{ttx}\|^2)(\tau)d\tau, \qquad \forall t > 0, \qquad (1.4.9)$$

$$\|\theta_{tt}(t)\|^2 + \int_0^t \|\theta_{ttx}(\tau)\| d\tau$$

$$\le C_4 + C_4 \int_0^t (\|\theta_{txx}\|^2 + \varepsilon\|v_{ttx}\|^2)(\tau)d\tau$$

$$+ C_4 \int_0^t (\|v_x\|^2 + \|v_{tx}\|^2 + \|\theta\|^2 + \|\theta_t\|^2)\|\theta_{tt}\|^2(\tau)d\tau, \qquad \forall t > 0. \quad (1.4.10)$$

Proof. Differentiating (1.1.10), (1.1.11) with respect to t, we can get

$$\|v_{tt}(t)\| \le C_4(\|\theta_x\| + \|u_x\| + \|v_{txx}\| + \|\theta_t\|_{H^1} + \|f_r\| + \|f_t\|), \qquad (1.4.11)$$

$$\|\theta_{tt}(t)\| \le C_4(\|\theta_t\|_{H^2} + \|v_x\| + \|v_{tx}\| + \|\theta_x\|_{H^1} + \|g_r\| + \|g_t\|). \qquad (1.4.12)$$

Differentiating (1.1.10) twice with respect to t, multiplying the resultant by v_{tt}, then integrating the result over $[0, L]$, using Young's inequality, we obtain

$$\frac{d}{dt}\|v_{tt}(t)\|^2 + C_1\|v_{ttx}(t)\|^2 \tag{1.4.13}$$

$$\leq C_4 \left(\|v_{xx}\|^2 + \|\theta_x\|^2 + \|v_{tt}\|^2 + \|v_{txx}\|^2 + \|\theta_{tx}\|^2 + \|\theta_{ttx}\|^2 + \left\|\frac{d^2 f}{dt^2}\right\|^2 \right).$$

Integrating (1.4.13) with respect to t, and noting that

$$\frac{d^2 f}{dt^2} = f_{rr}v^2 + 2f_{rt}v + f_r v_t + f_{tt},$$

then by (1.4.11), Theorems 1.2.1 and 1.3.1, and conditions (1.3.25), (1.3.1), we can get (1.4.9).

By the same method, differentiating (1.1.11) twice with respect to t, multiplying the resultant by θ_{tt}, then integrating the resultant over $[0, L]$, using Young's inequality, we have

$$\frac{d}{dt}\|\theta_{tt}(t)\|^2 \leq C_4 \Big\{ -\|\theta_{ttx}\|^2 + \varepsilon\|v_{ttx}\|^2 + \|\theta_{tt}\|^2 + \|v_{tx}\|^2 + \|\theta_t\|^2 + \|v_x\|^2$$

$$+ \|\theta\|^2 + \left\|\frac{d^2 g}{dt^2}\right\|^2 \Big\} + C_4 \left(\|v_x\|^2 + \|v_{tx}\|^2 + \|\theta\|^2 + \|\theta_t\|^2 \right) \|\theta_{tt}\|^2. \tag{1.4.14}$$

Integrating (1.4.14) with respect to t, using (1.4.12), conditions (1.4.2), (1.3.2), Theorems 1.2.1, 1.3.1 and noting that

$$\frac{d^2 g}{dt^2} = g_{rr}v^2 + 2g_{rt}v + g_r v_t + g_{tt},$$

we derive (1.4.10). $\qquad\square$

Lemma 1.4.2. *Under conditions of Theorem 1.4.1, we have that for any $t > 0$, and for any $\varepsilon > 0$,*

$$\|v_{tx}(t)\|^2 + \int_0^t \|v_{txx}(\tau)\|^2 d\tau \leq C_4 + C_4 \int_0^t \varepsilon\|\theta_{txx}\|^2(\tau)d\tau, \tag{1.4.15}$$

$$\|\theta_{tx}(t)\|^2 + \int_0^t \|\theta_{txx}(\tau)\|^2 d\tau$$

$$\leq C_4 + C_4 \int_0^t \left[\varepsilon\|v_{txx}\|^2 + (\|v_x\|^2 + \|\theta\|_{H^1}^2 + \|v_{xx}\|^2)\|\theta_{tx}\|^2 \right](\tau)d\tau. \tag{1.4.16}$$

Proof. Differentiating (1.1.10) with respect to t and x, multiplying by v_{tx}, then integrating the result with respect to x by parts, using Young's inequality, we have for any $\varepsilon > 0$,

$$\frac{d}{dt}\|v_{tx}(t)\|^2 + C_1\|v_{txx}(t)\|^2 \le C_4\Big(\|v_{xx}\|^2 + \|\theta_x\|_{H^1}^2 + \varepsilon\|v_{txx}\|^2 + \|v_{tx}\|^2$$

$$+ \|v_x\|_{H^2}^2 + \varepsilon\|\theta_{txx}\|^2 + \left\|\frac{d^2 f}{dtdx}\right\|^2\Big). \tag{1.4.17}$$

Integrating (1.4.17) with respect to t, noting that $\dfrac{d^2 f}{dtdx} = f_{rr}ur^{1-n}v + f_{rt}r^{1-n}u + f_r v_x$, then by Theorems 1.2.1–1.3.1 and conditions (1.4.1), (1.3.1), we have for $\varepsilon > 0$ small enough,

$$\|v_{tx}(t)\|^2 + \int_0^t \|v_{txx}(\tau)\|^2 d\tau \le C_4 + C_4\varepsilon \int_0^t \|\theta_{txx}(\tau)\|^2 d\tau.$$

By the same method, differentiating (1.1.11) with respect to t and x, multiplying the result by θ_{tx}, then integrating with respect to x, using Young's inequality, we obtain

$$\frac{d}{dt}\|\theta_{tx}(t)\|^2 \le C_4\Big(-\|\theta_{txx}\|^2 + \varepsilon\|v_{txx}\|^2 + \|v_{tx}\|^2 + \|\theta_t\|^2 + \left\|\frac{d^2 g}{dtdx}\right\|^2\Big)$$

$$+ C_4\left(\|v_x\|^2 + \|v_{xx}\|^2 + \|\theta_x\|_{H^1}^2\right)\|\theta_{tx}\|^2. \tag{1.4.18}$$

Integrating (1.4.18) with respect to t, then by Theorems 1.2.1–1.3.1 and conditions (1.4.2), (1.3.2), we derive

$$\|\theta_{tx}(t)\|^2 + \int_0^t \|\theta_{txx}(\tau)\|^2 d\tau$$

$$\le C_4 + C_4 \int_0^t \left[\varepsilon\|v_{txx}\|^2 + (\|v_x\|^2 + \|\theta_x\|_{H^1}^2 + \|v_{xx}\|^2)\|\theta_{tx}\|^2\right](\tau)d\tau. \qquad \square$$

Lemma 1.4.3. *Under the conditions of Theorem 1.4.1, we have that for any $t > 0$,*

$$\|v_{tt}(t)\|^2 + \|\theta_{tt}(t)\|^2 + \|v_{tx}(t)\|^2 + \|\theta_{tx}(t)\|^2$$

$$+ \int_0^t \left(\|v_{txx}\|^2 + \|\theta_{txx}\|^2 + \|v_{ttx}\|^2 + \|\theta_{ttx}\|^2\right)(\tau)d\tau \le C_4. \tag{1.4.19}$$

Proof. Adding (1.4.15) to (1.4.16), we have

$$\|v_{tx}(t)\|^2 + \|\theta_{tx}(t)\|^2 + \int_0^t \left(\|v_{txx}\|^2 + \|\theta_{txx}\|^2\right)(\tau)d\tau$$

$$\le C_4 + C_4 \int_0^t \left(\|v_x\|^2 + \|\theta_x\|_{H^1}^2 + \|v_{xx}\|^2\right)\|\theta_{tx}\|^2(\tau)d\tau. \tag{1.4.20}$$

Using Gronwall's inequality, and Theorems 1.2.1–1.3.1, we can derive

$$\|v_{tx}(t)\|^2 + \|\theta_{tx}(t)\|^2 + \int_0^t \left(\|v_{txx}\|^2 + \|\theta_{txx}\|^2 \right)(\tau)d\tau \le C_4. \qquad (1.4.21)$$

Multiplying (1.4.9), (1.4.10) by ε respectively, then adding the results up to (1.4.13), we obtain

$$\|v_{tt}(t)\|^2 + \|\theta_{tt}(t)\|^2 + \|v_{tx}(t)\|^2 + \|\theta_{tx}(t)\|^2$$

$$+ \int_0^t \left(\|v_{txx}\|^2 + \|\theta_{txx}\|^2 + \|v_{ttx}\|^2 + \|\theta_{ttx}\|^2 \right)(\tau)d\tau$$

$$\le C_4 + C_4(\varepsilon) \int_0^t \left(\|v_x\|^2 + \|v_{tx}\|^2 + \|\theta_x\|^2 + \|\theta_t\|^2 \right) \|\theta_{tt}\|^2(\tau)d\tau. \qquad (1.4.22)$$

Applying Gronwall's inequality to (1.4.22), and Theorems 1.2.1, 1.3.2, we get (1.4.19). $\qquad \square$

Lemma 1.4.4. *Under conditions of Theorem 1.4.1, it holds that for any $t > 0$,*

$$\|u_{xxx}(t)\|_{H^1}^2 + \|u_{xx}(t)\|_{W^{1,\infty}}^2$$

$$+ \int_0^t \left(\|u_{xxx}\|_{H^1}^2 + \|u_{xx}\|_{W^{1,\infty}}^2 \right)(\tau)d\tau \le C_4, \qquad (1.4.23)$$

$$\|v_{xxx}(t)\|_{H^1}^2 + \|v_{xx}(t)\|_{W^{1,\infty}}^2 + \|\theta_{xxx}(t)\|_{H^1}^2 + \|\theta_{xx}(t)\|_{W^{1,\infty}}^2 \qquad (1.4.24)$$

$$+ \|u_{txxx}(t)\|^2 + \|v_{txx}(t)\|^2 + \|\theta_{txx}(t)\|^2$$

$$+ \int_0^t (\|v_{tt}\|^2 + \|\theta_{tt}\|^2 + \|\theta_{txx}\|_{H^1}^2 + \|v_{txx}\|_{H^1}^2 + \|u_{txxx}\|_{H^1}^2)(\tau)d\tau \le C_4,$$

$$\int_0^t (\|v_{xxxx}\|_{H^1}^2 + \|\theta_{xxxx}\|_{H^1}^2)(\tau)d\tau \le C_4. \qquad (1.4.25)$$

Proof. Differentiating (1.1.10) with respect to x, and using equation (1.1.9), we get

$$\beta \frac{d}{dt} \left(\frac{u_{xxx}}{u} \right) + R \frac{\theta u_{xxx}}{u^2}$$

$$= (r^{1-n})_{xx} v_t + 2(r^{1-n})_x v_{tx} + r^{1-n} v_{txx} + \beta \left\{ 3 \frac{(r^{n-1}v)_{xxx} u_x}{u^2} + 3 \frac{(r^{n-1}v)_{xx} u_{xx}}{u^2} \right.$$

$$+ 6 \frac{(r^{n-1}v)_x u_x^3}{u^4} - 6 \frac{(r^{n-1}v)_{xx} u_x^2}{u^3} - 6 \frac{(r^{n-1}v)_x u_{xx} u_x^2}{u^3} \left. \right\} + R \left\{ \frac{\theta_{xxx}}{u} - 3 \frac{\theta_{xx} u_x}{u^2} \right.$$

$$- 3 \frac{\theta_x u_{xx}}{u^2} + 4 \frac{\theta_x u_x^2}{u^3} + 2 \frac{\theta u_{xx} u_x}{u^3} + 2 \frac{\theta_x u_x^2}{u^3} + 4 \frac{\theta u_x u_{xx}}{u^3} - 6 \frac{u_x^3 \theta}{u^4} \left. \right\} - (f_{rr} u^2 + f_r u_x)$$

$$:= E(x,t) \qquad (1.4.26)$$

with

$$\|E(t)\| \le C_4 \left(\|v_t\|_{H^2} + \|u_x\|_{H^1} + \|v_x\|_{H^2} + \|\theta_x\|_{H^2} + \|f_{rr}\| + \|f_r\| \right).$$

By Theorem 1.3.1, Lemma 1.4.3 and condition (1.4.1), we have

$$\int_0^t \|E(\tau)\|^2 d\tau \le C_4, \quad \forall t > 0. \tag{1.4.27}$$

Multiplying (1.4.26) by $\dfrac{u_{xxx}}{u}$, then integrating the resultant over $[0, L] \times [0, t]$, we obtain

$$\|u_{xxx}(t)\|^2 + \int_0^t \|u_{xxx}(\tau)\|^2 d\tau \le C_4. \tag{1.4.28}$$

Differentiating (1.1.10) with respect to x, by Theorem 1.3.1 and Lemma 1.4.3, we can deduce that

$$\|v_{xxx}(t)\| \le C_4 \left(\|v_{tx}\| + \|\theta\|_{H^2} + \|v\|_{H^2} + \|u\|_{H^2} + \|f_r u\| \right) \le C_4. \tag{1.4.29}$$

Differentiating (1.1.10) with respect to x twice, we have

$$\|v_{xxxx}(t)\| \le C_4 \left(\|v_{txx}\| + \|\theta\|_{H^3} + \|v\|_{H^3} + \|u\|_{H^3} + \|f_{rr}\| + \|f_r\| \right). \tag{1.4.30}$$

By the same method, differentiating (1.1.11) with respect to x once and twice respectively, we get

$$\|\theta_{xxx}(t)\| \le C_4 (\|\theta_{tx}\| + \|v\|_{H^2} + \|\theta\|_{H^2} + \|u\|_{H^2} + \|g_r\|) \le C_4, \tag{1.4.31}$$

$$\|\theta_{xxxx}(t)\| \le C_4 (\|\theta_{txx}\| + \|v\|_{H^3} + \|\theta\|_{H^3} + \|u\|_{H^3} + \|g_{rr}\| + \|g_r\|). \tag{1.4.32}$$

By Theorem 1.3.1 and (1.4.20), we have

$$\|v_{xxx}(t)\|^2 + \|\theta_{xxx}(t)\|^2 + \int_0^t \left(\|v_{xxx}\|_{H^1}^2 + \|\theta_{xxx}\|_{H^1}^2 \right)(\tau) d\tau \le C_4. \tag{1.4.33}$$

Differentiating (1.1.10), (1.1.11) with respect to t respectively, using Theorem 1.3.1 and Lemma 1.4.3, we can obtain

$$\|v_{txx}(t)\| \le C_4 \left(\|v_{tt}\| + \|\theta_{tx}\| + \|u_{tx}\| + \|f_r\| + \|f_t\| + \|v_x\|_{H^1} \right) \le C_4, \tag{1.4.34}$$

$$\|\theta_{txx}(t)\| \le C_4 \left(\|\theta_{tt}\| + \|\theta_t\| + \|v_{tx}\| + \|u_{tx}\| + \|g_r\| + \|g_t\| \right) \le C_4. \tag{1.4.35}$$

By (1.4.30), (1.4.32)–(1.4.35), Theorem 1.3.1 and conditions (1.4.1), (1.4.2), we have

$$\|v_{xxxx}(t)\|^2 + \|\theta_{xxxx}(t)\|^2 + \int_0^t (\|v_{xxxx}\|^2 + \|\theta_{xxxx}\|^2)(\tau) d\tau \le C_4. \tag{1.4.36}$$

Differentiating (1.4.26) with respect to x, we can get

$$\beta \frac{d}{dt}\left(\frac{u_{xxxx}}{u}\right) + R\frac{\theta u_{xxxx}}{u^2}$$

$$= \frac{(r^{n-1}v)_{xxxx}u_x}{u^2} + \frac{u_{xxx}(r^{n-1}v)_{xx}}{u^2} - 2\frac{u_{xxx}u_x^2}{u^3} + 2\frac{\theta u_{xxx}u_x}{u^3} - \frac{\theta_x u_{xxx}}{u^2} + E_x$$

$$= E_1(x,t) \tag{1.4.37}$$

with

$$\|E_1(t)\| \le C_4\left(\|u\|_{H^3} + \|v\|_{H^4} + \|\theta\|_{H^4} + \|v_t\|_{H^3} + \left\|\frac{d^3f}{dx^3}\right\|\right). \tag{1.4.38}$$

Differentiating (1.1.10) with respect to t and x, we can derive

$$\|v_{txxx}(t)\| \tag{1.4.39}$$
$$\le C_4\left(\|v_{ttx}\| + \|\theta_{txx}\| + \|f_r\| + \|f_{rt}\| + \|f_{rr}\| + \|v_t\|_{H^2} + \|v\|_{H^3} + \|\theta\|_{H^2}\right).$$

Multiplying (1.4.37) by $\dfrac{u_{xxxx}}{u}$, then integrating the resultant over $[0,L] \times [0,t]$, using Theorem 1.3.1, Lemma 1.4.3, (1.4.28), (1.4.36), (1.4.39) and (1.4.1)–(1.4.2), we obtain

$$\|u_{xxxx}(t)\|^2 + \int_0^t \|u_{xxxx}(\tau)\|^2 d\tau \le C_4, \quad \forall t > 0. \tag{1.4.40}$$

Differentiating (1.1.10) and (1.1.11) three times with respect to x respectively, we get

$$\|v_{xxxxx}(t)\| \le C_4\left(\|v_t\|_{H^3} + \|v\|_{H^4} + \|\theta\|_{H^4} + \|u\|_{H^4} + \left\|\frac{d^3f}{dx^3}\right\|\right), \tag{1.4.41}$$

$$\|\theta_{xxxxx}(t)\| \le C_4\left(\|\theta_t\|_{H^3} + \|v\|_{H^4} + \|\theta\|_{H^4} + \|u\|_{H^3} + \left\|\frac{d^3g}{dx^3}\right\|\right). \tag{1.4.42}$$

Differentiating (1.1.11) with respect to t and x, we have

$$\|\theta_{txxx}(t)\| \tag{1.4.43}$$
$$\le C_4\left(\|\theta_{ttx}\| + \|\theta\|_{H^3} + \|\theta_t\|_{H^2} + \|v_t\|_{H^2} + \|v\|_{H^2} + \|u_{tx}\| + \left\|\frac{d^2g}{dtdx}\right\|\right).$$

By (1.1.9), we obtain

$$\|u_{txxx}(t)\| \le C_4\|v(t)\|_{H^4}, \qquad \|u_{txxx}(t)\|_{H^1} \le C_4\|v(t)\|_{H^5}. \tag{1.4.44}$$

Combining (1.4.28) and (1.4.40), by the embedding theorem, we get (1.4.23). Combining (1.4.41) and (1.4.42), by (1.4.36), (1.4.39), (1.4.40), (1.4.43) and conditions (1.4.1), (1.4.2), we obtain (1.4.25). By (1.4.33)–(1.4.36), (1.4.11), (1.4.12), (1.4.39), (1.4.43), (1.4.44) and the embedding theorem, we obtain (1.4.24). $\qquad\square$

Remark 1.4.2. The results in this chapter indicate that suitable regularities of the non-autonomous terms can ensure the global existence of solutions in H^i ($i = 1, 2, 4$).

Remark 1.4.3. Based on the uniform estimates in this chapter, we can established the asymptotic behavior of global solutions in H^i ($i = 1, 2, 4$), and even exponential stability of global solutions if we impose some proper exponential decay on non-autonomous terms f and g.

Proof of Theorem 1.4.1. By Lemmas 1.4.1–1.4.4, and noting the following estimates:

$$\|u_{ttx}(t)\| \le C_4\|v_{ttx}(t)\|, \quad \|u_{ttxx}(t)\| \le C_4\|v_{txxx}(t)\|, \quad \|u_{ttt}(t)\| \le C_4\|v_{ttx}(t)\|,$$

we can obtain Theorem 1.4.1.   □

1.5 Bibliographic Comments

For the case $f = g \equiv 0$, Fujita-Yashima and Benabidallah [80, 81] established the global existence of solutions to problem (1.1.9)–(1.1.15), Jiang [29] proved the large-time behavior of global solutions in H^1, Zheng and Qin [84] obtained the global existence of universal attractors in H^1 and H^2, Qin et al. [65] established the exponential stability of global solutions in H^4. We also refer the reader to the related results in Cho, Choe and Kim [6], Qin and Muñoz Rivera [71], Xu and Yang [78], Yanagi [79], Zheng [82], Zheng and Qin [83] and Zimmer [86].

Chapter 2

Global Existence and Exponential Stability for a Real Viscous Heat-conducting Flow with Shear Viscosity

2.1 Introduction

In this chapter we shall study the global existence and exponential stability of weak solutions for a real viscous compressible heat-conducting flow between two horizontal plates. The system describing this type of flow is derived from the following general $3D$ Navier-Stokes equations:

$$\rho_t + \operatorname{div}(\rho\vec{u}) = 0, \tag{2.1.1}$$

$$(\rho\vec{u})_t + \operatorname{div}(\rho\vec{u} \otimes \vec{u}) + \nabla p = \operatorname{div}\left(\lambda'(\operatorname{div}\vec{u})\right)\vec{I} + \mu(\nabla\vec{u} + (\nabla\vec{u})^T), \tag{2.1.2}$$

$$\zeta_t + \operatorname{div}\left(\vec{u}(\zeta + p)\right) = \operatorname{div}\left(\lambda'(\operatorname{div}\vec{u})\vec{u} + \mu\vec{u}(\nabla\vec{u} + (\nabla\vec{u})^T) + \kappa\nabla\theta\right) \tag{2.1.3}$$

where $x \in \mathbb{R}^3$ is the spatial variable and $t > 0$ is the time, $\rho \in \mathbb{R}, \vec{u} \in \mathbb{R}^3$ and $\theta \in \mathbb{R}^+$ denote the density, velocity and temperature, respectively, the total energy is $\zeta = \rho\left(e + \frac{1}{2}|\vec{u}|^2\right)$, with e the internal energy and $\frac{1}{2}|\vec{u}|^2$ the kinetic energy, the equations of state $p = p(\rho, \theta)$, $e = e(\rho, \theta)$ relate this pressure p and the internal energy e with the density and temperature of the flow, $(\nabla\vec{u})^T$ is the transpose of the matrix $\nabla\vec{u}$, $\lambda' = \lambda'(\rho, \theta)$ and $\mu = \mu(\rho, \theta)$ are the viscosity coefficients of the flow and $\kappa = \kappa(\rho, \theta)$ is the heat conductivity. The fluid in question is a Newtonian fluid, i.e., the stress tensor $-p\vec{I} + \lambda'(\operatorname{div}\vec{v})\vec{I} + \mu(\nabla\vec{u} + (\nabla\vec{u})^T)$ is a linear function of the deformation tensor $\frac{1}{2}(\nabla\vec{u} + (\nabla\vec{u})^T)$. The viscosity and heat conduction terms describe the dissipative processes in viscous flows.

Consider a $3D$ flow (2.1.1)–(2.1.3) with spatial variable $\vec{x} = (x, x_2, x_3)$, which is moving in the x direction and uniform in the transverse direction (x_2, x_3),

$$\rho = \rho(x, t), \quad \theta = \theta(x, t), \quad \vec{u} = (u, \vec{w})(x, t), \quad \vec{w} = (u_2, u_3), \tag{2.1.4}$$

where $u \in \mathbb{R}$ is the longitudinal velocity and $\vec{w} \in \mathbb{R}^2$ is the transverse velocity. With this structure (2.1.4), equations (2.1.1)–(2.1.3) can be written as the following in one space dimension with $\lambda = \lambda' + 2\mu > 0$:

$$\rho_t + (\rho u)_x = 0, \tag{2.1.5}$$

$$(\rho u)_t + (\rho u^2 + p)_x = (\lambda u_x)_x, \tag{2.1.6}$$

$$(\rho \vec{w})_t + (\rho u \vec{w})_x = (\mu \vec{w}_x)_x, \tag{2.1.7}$$

$$\zeta_t + (u(\zeta + p))_x = (\lambda u u_x + \mu \vec{w} \cdot \vec{w}_x + \kappa \theta_x)_x, \tag{2.1.8}$$

where as in (2.1.1)–(2.1.3), $x \in \mathbb{R}$ is the spatial variable and $t > 0$ is the time variable, $\rho \in \mathbb{R}^+$, $u \in \mathbb{R}$, $\vec{w} \in \mathbb{R}^2$, $\theta \in \mathbb{R}$, p denote the density, longitudinal velocity, transverse velocity, temperature, pressure respectively and the total energy of the plane viscous flow is

$$\zeta = \rho \left(e + \frac{1}{2}(u^2 + |\vec{w}|^2) \right), \tag{2.1.9}$$

with internal energy e. The pressure p and internal energy e are expressed by the density and the temperature of the flow according to the state equations

$$p = p(\rho, \theta), \quad e = e(\rho, \theta) \tag{2.1.10}$$

where $\lambda = \lambda(\rho, \theta)$ and $\mu = \mu(\rho, \theta)$ are the viscosity coefficients of the flow and $\kappa = \kappa(\rho, \theta)$ is the heat conductivity; $\mu = \mu(\rho, \theta)$ is particularly called the shear viscosity.

Consider the initial boundary value problem (2.1.5)–(2.1.8) in a bounded spatial domain $\Omega = (0, 1)$ with the following initial condition and boundary conditions:

$$(\rho, u, \vec{w}, \theta)|_{t=0} = (\rho_0, u_0, \vec{w}_0, \theta_0)(x), \quad x \in \Omega, \tag{2.1.11}$$

$$(u, \vec{w})\big|_{\partial\Omega} = 0, \ \theta_x\big|_{\partial\Omega} = 0, \ or \ \theta\big|_{\partial\Omega} = T_0 = \text{const.} > 0, \tag{2.1.12}$$

where the initial data are compatible with each other.

We assume that $\theta_0(x) > 0$, $\rho_0(x) > 0$, for any $x \in (0, 1)$,

$$\int_0^1 \rho_0(x)dx = 1. \tag{2.1.13}$$

Now we introduce the Lagrangian variable,

$$y = y(x, t) = \int_0^x \rho(\xi, t)d\xi. \tag{2.1.14}$$

Then we have from (2.1.5), (2.1.13) and (2.1.14),

$$0 \le y \le 1, \ \int_0^1 \rho(x, t)dx = \int_0^1 \rho_0(x)dx = 1. \tag{2.1.15}$$

Therefore we can translate the problem (2.1.1)–(2.1.3) in Euler coordinates into the following initial boundary value problem in Lagrangian coordinates (y, t), $y \in \Omega = (0, 1)$, a moving of coordinates along the particle path,

$$v_t - u_y = 0, \tag{2.1.16}$$

$$u_t + p_y = \left(\lambda \frac{u_y}{v}\right)_y, \tag{2.1.17}$$

$$\vec{w}_t = \left(\mu \frac{\vec{w}_y}{v}\right)_y, \tag{2.1.18}$$

$$E_t + (up)_y = \left(\frac{\lambda u u_y + \mu \vec{w}\vec{w}_y + \kappa \theta_y}{v}\right)_y, \tag{2.1.19}$$

with the initial boundary conditions

$$(v, u, \vec{w}, \theta)\big|_{t=0} = (v_0, u_0, \vec{w}_0, \theta_0)(y), \quad y \in \Omega, \tag{2.1.20}$$

$$(u, \vec{w}) = 0, \quad \theta_y = 0, \quad \text{on } \partial\Omega \times [0, +\infty), \tag{2.1.21}$$

or

$$(u, \vec{w}) = 0, \quad \theta_y = T_0 > 0, \quad \text{on } \partial\Omega \times [0, +\infty), \tag{2.1.22}$$

where $v = \dfrac{1}{\rho}$ is specific volume, $e = e(v, \theta), p = p(v, \theta)$, and

$$E = e + \frac{1}{2}(u^2 + |\vec{w}|^2). \tag{2.1.23}$$

The second law of thermodynamics states the relation between p and e,

$$e_v(v, \theta) + p(v, \theta) = \theta p_\theta(v, \theta). \tag{2.1.24}$$

Now we assume that e, p and κ are C^3 functions on $0 < u < +\infty$ and $0 \leq \theta < +\infty$. Let q and r be two positive constants (exponents of growth) satisfying one of the following relations:

$$0 \leq r \leq 1/3, \quad 1/3 < q, \tag{2.1.25}$$
$$1/3 < r < 4/7, \quad (2r+1)/5 < q, \tag{2.1.26}$$
$$4/7 \leq r \leq 1, \quad (5r+1)/9 < q, \tag{2.1.27}$$
$$1 < r \leq 13/3, \quad (9r+1)/15 < q, \tag{2.1.28}$$
$$13/3 < r, \quad (11r+3)/19 < q. \tag{2.1.29}$$

Moreover, we further assume that there are positive constants ν, p_0, p_1, k_0 and for any $\underline{v} > 0$, there are positive constants $N(\underline{v})$, $p_2(\underline{v})$, $p_3(\underline{v})$ and $k_1(\underline{v})$ such that for any $v \geq \underline{v}$ and $\theta \geq 0$ the following conditions hold:

$$0 \leq e(v, 0), \quad \nu(1 + \theta^r) \leq e_\theta(v, \theta) \leq N(\underline{v})(1 + \theta^r), \tag{2.1.30}$$

$$p_0\theta^{r+1} < vp(v, \theta) \leq p_1(1 + \theta^{r+1}), \tag{2.1.31}$$

$$- p_2(\underline{v})[l + (1-l)\theta + \theta^{r+1}] \leq p_v(v,\theta)$$
$$\leq -p_3(\underline{v})[l + (1-l)\theta + \theta^{r+1}], l = 0 \ \ or \ 1, \tag{2.1.32}$$
$$|p_\theta(v,\theta)| \leq p_3(\underline{v})(1 + \theta^r), \tag{2.1.33}$$
$$k_0(1 + \theta^q) \leq k(v,\theta) \leq k_1(\underline{v})(1 + \theta^q), \tag{2.1.34}$$
$$|k_v(v,\theta)| + |k_{vv}(v,\theta)| \leq k_1(\underline{v})(1 + \theta^q). \tag{2.1.35}$$

For the viscosity $\lambda(v,\theta)$, we assume that

$$\lambda(v,\theta) = \lambda_0 > 0 \tag{2.1.36}$$

is a constant.

The notation in the chapter is standard. We put $\| \cdot \| = \| \cdot \|_{L^2}$ and denote by $C^k(I, B)$, $k \in \mathbb{N}_0 \equiv \{0\} \cup \mathbb{N}$, the space of k-times continuously differentiable functions from $I \subseteq \mathbb{R}$ into a Banach space B, and likewise by $L^{\bar{p}}(I, B)$, $1 \leq \bar{p} \leq +\infty$, the corresponding Lebesgue spaces. Subscripts t, x and y denote the (partial) derivatives with respect to t, x and y, respectively. We use C_i $(i = 1, 2, 3, 4)$ to denote the universal positive constants depending only on the H^i norm of initial data, $\min_{y \in [0,1]} v_0(y)$, $\min_{y \in [0,1]} \vec{w}_0(y)$ and $\min_{y \in [0,1]} \theta_0(y)$. Without danger of confusion, we will use the same symbol to denote the state functions as well as their values along the thermodynamic process, e.g., $p(u, \theta)$ and $p(u(x,t), \theta(x,t))$ and $p(x,t)$.

Define the following three spaces:

$$H_+^1 = \Big\{ (v, u, \vec{w}, \theta) \in (H^1[0,1])^5 : v(x) > 0, \theta(x) > 0, \forall x \in [0,1],$$
$$u(0) = u(1) = 0, \vec{w}(0) = \vec{0}, \ \theta(0) = \theta(1) = T_0 \ for \ (2.1.22) \Big\},$$

$$H_+^i = \Big\{ (v, u, \vec{w}, \theta) \in (H^i[0,1])^5 : v(x) > 0, \theta(x) > 0, \forall x \in [0,1],$$
$$u(0) = u(1) = 0, \vec{w}(0) = \vec{0}, \ \theta(0) = \theta(1) = T_0 \ for \ (2.1.22),$$
$$\theta'(0) = \theta'(1) = 0 \ for \ (2.1.21) \Big\}, \ i = 2, 4.$$

Our results read as follows, which are selected from [60, 64].

Theorem 2.1.1. *Assume that e, p and κ are C^2 functions on $0 < v < +\infty$ and $0 \leq \theta < +\infty$, and assumptions (2.1.23)–(2.1.36) hold. Then for any $(v_0, u_0, \vec{w}_0, \theta_0) \in H_+^1$, there exists a unique global solution $(v(t), u(t), \vec{w}(t), \theta(t)) \in H_+^1$ to problem (2.1.16)–(2.1.21) or (2.1.16)–(2.1.20), (2.1.22) verifying that,*

$$0 < C_1^{-1} \leq \theta(y,t) \leq C_1, \quad 0 < C_1^{-1} \leq u(y,t) \leq C_1, \quad \forall (y,t) \in [0,1] \times [0,+\infty) \tag{2.1.37}$$

and for any $t > 0$,

$$\|v(t) - \bar{v}\|_{H^1}^2 + \|u(t)\|_{H^1}^2 + \|\vec{w}(t)\|_{H^1}^2 + \|\theta(t) - \bar{\theta}\|_{H^1}^2$$

$$+ \int_0^t \left(\|v - \bar{v}\|_{H^1}^2 + \|u\|_{H^2}^2 + \|\vec{w}\|_{H^2}^2 + \|\theta - \bar{\theta}\|_{H^2}^2 + \|u_t\|^2 \right.$$

$$\left. + \|\vec{w}_t\|^2 + \|\theta_t\|^2 \right)(\tau) \, d\tau \leq C_1. \tag{2.1.38}$$

Moreover, there exist constants $C_1 > 0$ and $\gamma_1 = \gamma_1(C_1) > 0$ such that and for any $t > 0$,

$$e^{\gamma t} \left(\|v(t) - \bar{v}\|_{H^1}^2 + \|u(t)\|_{H^1}^2 + \|\vec{w}(t)\|_{H^1}^2 + \|\theta(t) - \bar{\theta}\|_{H^1}^2 \right)$$

$$+ \int_0^t e^{\gamma \tau} \left(\|v - \bar{v}\|_{H^1}^2 + \|u\|_{H^2}^2 + \|\vec{w}\|_{H^2}^2 + \|\theta - \bar{\theta}\|_{H^2}^2 \right.$$

$$\left. + \|u_t\|^2 + \|\vec{w}_t\|^2 + \|\theta_t\|^2 \right)(\tau) \, d\tau \leq C_1 \tag{2.1.39}$$

where $\bar{v} = \int_0^1 v_0(x) \, dx$, $\bar{\theta} = T_0$ for (2.1.22), $e(\bar{v}, \bar{\theta}) = \int_0^1 (e(v_0, \theta_0) + \frac{v_0^2}{2})(x) \, dx$ for (2.1.21).

Theorem 2.1.2. *Assume that e, p and κ are C^3 functions on $0 < v < +\infty$, $0 \leq \theta < +\infty$, and assumptions (2.1.23)–(2.1.36) hold. Then for any $(v_0, u_0, \vec{w}_0, \theta_0) \in H_+^2$, there exists a unique global solution $(v(t), u(t), \vec{w}(t), \theta(t)) \in H_+^2$ to problem (2.1.16)–(2.1.21) or (2.1.16)–(2.1.20), (2.1.22) verifying that for any $t > 0$,*

$$\|v(t) - \bar{v}\|_{H^2}^2 + \|u(t)\|_{H^2}^2 + \|\vec{w}(t)\|_{H^2}^2 + \|\theta(t) - \bar{\theta}\|_{H^2}^2 + \|u_t(t)\|^2$$

$$+ \|\vec{w}_t(t)\|^2 + \|\theta_t(t)\|^2 + \int_0^t \left(\|v - \bar{v}\|_{H^2}^2 + \|u\|_{H^3}^2 + \|\vec{w}\|_{H^3}^2 \right.$$

$$\left. + \|\theta - \bar{\theta}\|_{H^3}^2 + \|u_{ty}\|^2 + \|\vec{w}_{ty}\|^2 + \|\theta_{ty}\|^2 \right)(\tau) \, d\tau \leq C_2, \tag{2.1.40}$$

and there exist constants $C_2 > 0$ and $\gamma_2 = \gamma_2(C_2) > 0$ such that for any fixed $\gamma \in (0, \gamma_2]$, the following estimates hold for any $t > 0$:

$$e^{\gamma t} \left(\|v(t) - \bar{v}\|_{H^2}^2 + \|u(t)\|_{H^2}^2 + \|\vec{w}(t)\|_{H^2}^2 + \|\theta(t) - \bar{\theta}\|_{H^2}^2 + \|u_t(t)\|^2 \right.$$

$$\left. + \|\vec{w}_t(t)\|^2 + \|\theta_t(t)\|^2 \right) + \int_0^t e^{\gamma t} \left(\|v - \bar{v}\|_{H^2}^2 + \|u\|_{H^3}^2 + \|\vec{w}\|_{H^3}^2 \right.$$

$$\left. + \|\theta - \bar{\theta}\|_{H^3}^2 + \|u_{ty}\|^2 + \|\vec{w}_{ty}\|^2 + \|\theta_{ty}\|^2 \right)(\tau) d\tau \leq C_2. \tag{2.1.41}$$

Theorem 2.1.3. *Assume that e, p are C^5 functions on $0 < v < +\infty$ and $0 \leq \theta < +\infty$, and assumptions (2.1.23)–(2.1.36) hold. Then for any $(v_0, u_0, \vec{w}_0, \theta_0) \in H_+^4$, there exists a unique global solution $(v(t), u(t), \vec{w}(t), \theta(t)) \in H_+^4$ to problem*

(2.1.16)–(2.1.21) *or* (2.1.16)–(2.1.20), (2.1.22) *verifying that for any* $t > 0$,

$$\|v(t) - \bar{v}\|^2_{H^4} + \|v(t) - \bar{v}\|^2_{W^{3,\infty}} + \|v_t(t)\|^2_{H^3} + \|v_{tt}(t)\|^2_{H^1} + \|u(t)\|^2_{H^4}$$
$$+ \|u(t)\|^2_{W^{3,\infty}} + \|u_t(t)\|^2_{H^2} + \|u_{tt}(t)\|^2 + \|\theta(t) - \bar{\theta}\|^2_{H^4}$$
$$+ \|\theta(t) - \bar{\theta}\|^2_{W^{3,\infty}} + \|\theta_t(t)\|^2_{H^2} + \|\theta_{tt}(t)\|^2 + \|\vec{w}(t)\|^2_{H^4}$$
$$+ \|\vec{w}(t)\|^2_{W^{3,\infty}} + \|\vec{w}_t(t)\|^2_{H^2} + \|\vec{w}_{tt}(t)\|^2 \le C_4, \tag{2.1.42}$$

$$\int_0^t \Big(\|v - \bar{v}\|^2_{H^4} + \|v - \bar{v}\|^2_{W^{3,\infty}} + \|v_t\|^2_{H^4} + \|v_{tt}\|^2_{H^2} + \|v_{ttt}\|^2$$
$$+ \|u\|^2_{H^5} + \|u\|^2_{W^{4,\infty}} + \|u_t\|^2_{H^3} + \|u_{tt}\|^2_{H^1} + \|\theta - \bar{\theta}\|^2_{H^5}$$
$$+ \|\theta - \bar{\theta}\|^2_{W^{4,\infty}} + \|\theta_t\|^2_{H^3} + \|\theta_{tt}\|^2_{H^1} + \|\vec{w}\|^2_{H^5} + \|\vec{w}\|^2_{W^{4,\infty}}$$
$$+ \|\vec{w}_t\|^2_{H^3} + \|\vec{w}_{tt}\|^2_{H^1} \Big)(\tau)\,d\tau \le C_4, \tag{2.1.43}$$

and there exist constants $C_4 > 0$ *and* $\gamma_4 = \gamma_4(C_4) > 0$ *such that for any fixed* $\gamma \in (0, \gamma_4]$, *the following estimates hold for any* $t > 0$:

$$\|v(t) - \bar{v}\|^2_{H^4} + \|v(t) - \bar{v}\|^2_{W^{3,\infty}} + \|v_t(t)\|^2_{H^3} + \|v_{tt}(t)\|^2_{H^1} + \|u(t)\|^2_{H^4}$$
$$+ \|u(t)\|^2_{W^{3,\infty}} + \|u_t(t)\|^2_{H^2} + \|u_{tt}(t)\|^2 + \|\theta(t) - \bar{\theta}\|^2_{H^4}$$
$$+ \|\theta(t) - \bar{\theta}\|^2_{W^{3,\infty}} + \|\theta_t(t)\|^2_{H^2} + \|\theta_{tt}(t)\|^2 + \|\vec{w}(t)\|^2_{H^4}$$
$$+ \|\vec{w}(t)\|^2_{W^{3,\infty}} + \|\vec{w}_t(t)\|^2_{H^2} + \|\vec{w}_{tt}(t)\|^2 \le C_4 e^{-\gamma t}, \tag{2.1.44}$$

$$\int_0^t e^{\gamma \tau} \Big(\|v - \bar{v}\|^2_{H^4} + \|v - \bar{v}\|^2_{W^{3,\infty}} + \|v_t\|^2_{H^4} + \|v_{tt}\|^2_{H^2} + \|v_{ttt}\|^2$$
$$+ \|u\|^2_{H^5} + \|u\|^2_{W^{4,\infty}} + \|u_t\|^2_{H^3} + \|u_{tt}\|^2_{H^1} + \|\theta - \bar{\theta}\|^2_{H^5}$$
$$+ \|\theta - \bar{\theta}\|^2_{W^{4,\infty}} + \|\theta_t\|^2_{H^3} + \|\theta_{tt}\|^2_{H^1} + \|\vec{w}\|^2_{H^5} + \|\vec{w}\|^2_{W^{4,\infty}}$$
$$+ \|\vec{w}_t\|^2_{H^3} + \|\vec{w}_{tt}\|^2_{H^1} \Big)(\tau)\,d\tau \le C_4. \tag{2.1.45}$$

2.2   Proof of Theorem 2.1.1

In this section, we shall complete the proof of Theorem 2.1.1 and take that the assumptions in Theorem 2.1.1 to be valid. We begin with the following lemma.

Lemma 2.2.1. *Assume that* e, p *and* κ *are* C^2 *functions on* $0 < v < +\infty$ *and* $0 \le \theta < +\infty$, *and assumptions* (2.1.23)–(2.1.36) *hold. Then for any* $(v_0, u_0, \vec{w}_0, \theta_0) \in H^1_+$, *there exists a unique global solution* $(v(t), u(t), \vec{w}(t), \theta(t)) \in H^1_+$ *to problem* (2.1.16)–(2.1.21) *or* (2.1.16)–(2.1.20), (2.1.22) *verifying that,*

$$\|v(t) - \bar{v}\|^2_{H^1} + \|u(t)\|^2_{H^1} + \|\vec{w}(t)\|^2_{H^1} + \|\theta(t) - \bar{\theta}\|^2_{H^1} \tag{2.2.1}$$
$$+ \int_0^t \Big(\|v - \bar{v}\|^2_{H^1} + \|u\|^2_{H^2} + \|\vec{w}\|^2_{H^2} + \|\theta - \bar{\theta}\|^2_{H^2} \Big)(\tau)\,d\tau \le C_1, \quad \forall t > 0$$

and for any $(y, t) \in [0, 1] \times [0, +\infty)$,

$$0 < C_1^{-1} \le \theta(y, t) \le C_1, \quad 0 < C_1^{-1} \le v(y, t) \le C_1. \tag{2.2.2}$$

Proof. For the case $\vec{w} = 0$, Qin [55] proved the result to problems (2.1.16), (2.1.17), (2.1.19)–(2.1.21) or (2.1.16), (2.1.17), (2.1.19)–(2.1.20), (2.1.22) under the assumptions (2.1.23)–(2.1.36). The proof is the same as that in Qin [50, 51, 52, 54] and Chapter 3 of [59] for the case of $\vec{w} = 0$. The proof is complete. $\square$

In what follows we shall prove the exponential stability in H_+^1. Set

$$\Psi(v, \theta) = e(v, \theta) - \theta \eta(\theta, v) \tag{2.2.3}$$

where $\eta(\theta, v)$ is defined by the relations

$$e_\theta = \theta \eta_\theta, \quad \eta_v = p_\theta. \tag{2.2.4}$$

Now we introduce the density of the gas, $\rho = 1/v$, then $\eta = \eta(1/\rho, \theta)$ satisfies

$$\frac{\partial \eta}{\partial \rho} = \frac{-p_\theta}{\rho^2}, \quad \frac{\partial \eta}{\partial \theta} = \frac{e_\theta}{\theta}. \tag{2.2.5}$$

We consider the transform

$$A : (\rho, \theta) \in D_{\rho,\theta} = \{(\rho, \theta) : \rho > 0, \theta > 0\} \to (v, \eta) \in AD_{\rho,\theta}. \tag{2.2.6}$$

Owing to the Jacobian, $|\partial(v, \eta)/\partial(\rho, \theta)| = -e_\theta/\rho^2\theta < 0$ on $AD_{\rho,\theta}$. Thus the functions e, p can also be regarded as the smooth functions of (v, η). We denote them by

$$e = e(v, \eta) :\equiv e(v, \theta(v, \eta)) = e(1/\rho, \theta),$$
$$p = p(v, \eta) :\equiv p(v, \theta(v, \eta)) = p(1/\rho, \theta). \tag{2.2.7}$$

Then it follows from (2.2.3)–(2.2.7) that

$$e_v = -p, \quad e_\eta = \theta,$$
$$p_v = -(\rho^2 p_\rho + \theta p_\theta^2/e_\theta), \quad p_\eta = \theta p_\theta/e_\theta,$$
$$\theta_v = -\theta p_\theta/e_\theta, \quad \theta_\eta = \theta/e_\theta. \tag{2.2.8}$$

We define the energy form

$$V(v, u, \vec{w}, \eta) = \frac{u^2}{2} + \frac{|\vec{w}|^2}{2} + e(v, \eta) - e(\bar{v}, \bar{\eta}) - \frac{\partial e}{\partial v}(\bar{v}, \bar{\eta})(v - \bar{v}) - \frac{\partial e}{\partial \eta}(\bar{v}, \bar{\eta})(\eta - \bar{\eta}), \tag{2.2.9}$$

where

$$\bar{v} = 1/\bar{\rho}, \quad \bar{\eta} = \eta(1/\bar{\rho}, \bar{\theta}), \tag{2.2.10}$$

$$\bar{\theta} = T_0, \text{ for } (2.1.22), \quad e(\bar{v}, \bar{\theta}) = \int_0^1 (e(v_0, \theta_0) + \frac{v_0^2}{2})(x) \, dx \text{ for } (2.1.21). \quad \square$$

Lemma 2.2.2. *The unique global solution* $(v(t), u(t), \vec{w}(t), \theta(t)) \in H^1_+$ *to problem* (2.1.16)–(2.1.21) *or* (2.1.16)–(2.1.20), (2.1.22) *satisfies the following estimates:*

$$\frac{u^2}{2} + \frac{|\vec{w}|^2}{2} + C_1^{-1} \left[(v - \bar{v})^2 + (\eta - \bar{\eta})^2 \right] \le V(v, u, \vec{w}, \eta) \tag{2.2.11}$$

$$\le \frac{u^2}{2} + \frac{|\vec{w}|^2}{2} + C_1 \left[(v - \bar{v})^2 + (\eta - \bar{\eta})^2 \right].$$

Proof. By the mean value problem theorem, there exists a point $(\tilde{v}, \tilde{\eta})$ between (v, η) and $(\bar{v}, \bar{\eta})$ such that

$$e(v, \eta) = e(\bar{v}, \bar{\eta}) + \frac{\partial e}{\partial v}(\bar{v}, \bar{\eta})(v - \bar{v}) + \frac{\partial e}{\partial \eta}(\bar{v}, \bar{\eta})(\eta - \bar{\eta}) + \frac{1}{2} \left[\frac{\partial^2 e}{\partial v^2}(\tilde{v}, \tilde{\eta})(v - \bar{v})^2 \right.$$

$$\left. + 2 \frac{\partial^2 e}{\partial v \partial \eta}(\tilde{v}, \tilde{\eta})(v - \bar{v})(\eta - \bar{\eta}) + \frac{\partial^2 e}{\partial \eta^2}(\tilde{v}, \tilde{\eta})(\eta - \bar{\eta})^2 \right]. \tag{2.2.12}$$

By (2.2.9) and (2.2.12), we get

$$V(v, u, \vec{w}, \eta) = \frac{u^2}{2} + \frac{|\vec{w}|^2}{2} + \frac{1}{2} \left[\frac{\partial^2 e}{\partial v^2}(\tilde{v}, \tilde{\eta})(v - \bar{v})^2 \right.$$

$$\left. + 2 \frac{\partial^2 e}{\partial v \partial \eta}(\tilde{v}, \tilde{\eta})(v - \bar{v})(\eta - \bar{\eta}) + \frac{\partial^2 e}{\partial \eta^2}(\tilde{v}, \tilde{\eta})(\eta - \bar{\eta})^2 \right] \tag{2.2.13}$$

where

$$\tilde{v} = \lambda_0 \bar{v} + (1 - \lambda_0) v, \quad \tilde{\eta} = \lambda_0 \bar{\eta} + (1 - \lambda_0) \eta, \quad 0 \le \lambda_0 \le 1.$$

By (2.2.2), we get

$$0 < C_1^{-1} \le \left| \tilde{v}(1/\tilde{\rho}, \tilde{\theta}) \right| \le C_1, \quad 0 < C_1^{-1} \le \left| \tilde{\eta}(1/\tilde{\rho}, \tilde{\theta}) \right| \le C_1,$$

which implies

$$\left| \frac{\partial^2 e}{\partial v^2}(\tilde{v}, \tilde{\eta}) \right| + \left| \frac{\partial^2 e}{\partial v \partial \eta}(\tilde{v}, \tilde{\eta}) \right| + \left| \frac{\partial^2 e}{\partial \eta^2}(\tilde{v}, \tilde{\eta}) \right| \le C_1. \tag{2.2.14}$$

Thus (2.2.14) and the Cauchy inequality give

$$V(u, v, \vec{w}, \eta) \le \frac{u^2}{2} + \frac{|\vec{w}|^2}{2} + C_1 \left[(v - \bar{v})^2 + (\eta - \bar{\eta})^2 \right]. \tag{2.2.15}$$

On the other hand, we infer from (2.2.8) that

$$e_{vv} = -p_v = \rho^2 p_\rho + \theta p_\theta^2 / e_\theta, \quad e_{v\eta} = -p_\eta = \theta_v = -\theta p_\theta / e_\theta, \quad e_{\eta\eta} = \theta_\eta = \theta / e_\theta,$$

which yields that the Hessian of $e(v, \eta)$ is positive definite for any $v > 0$ and $\theta > 0$. Thus we deduce from (2.2.12) that

$$V(v, u, \vec{w}, \eta) \ge \frac{u^2}{2} + \frac{|\vec{w}|^2}{2} + \lambda_{\min}(\tilde{v}, \tilde{\eta}) \left[(v - \bar{v})^2 + (\eta - \bar{\eta})^2 \right]$$

$$\ge \frac{u^2}{2} + \frac{|\vec{w}|^2}{2} + C_1^{-1} \left[(v - \bar{v})^2 + (\eta - \bar{\eta})^2 \right], \tag{2.2.16}$$

where $\lambda_{\min}(\tilde{v}, \tilde{\eta})$ $(\geq C_1^{-1})$ is the smaller characteristic root of the Hessian of $e(\tilde{v}, \tilde{\eta})$. Thus by the combination of (2.2.15) and (2.2.16), we deduce that

$$\frac{u^2}{2} + \frac{|\vec{w}|^2}{2} + C_1^{-1}\left[(v - \bar{v})^2 + (\eta - \bar{\eta})^2\right] \leq V(v, u, \vec{w}, \eta)$$

$$\leq \frac{u^2}{2} + \frac{|\vec{w}|^2}{2} + C_1\left[(v - \bar{v})^2 + (\eta - \bar{\eta})^2\right].$$

The proof is complete. $\qquad\qquad\square$

Lemma 2.2.3. *There exist constants $C_1 > 0$ and $\gamma_1 = \gamma_1(C_1) > 0$ such that for any fixed $\gamma \in (0, \gamma_1]$, the global solution $(v(t), u(t), \vec{w}(t), \theta(t)) \in H_+^1$ to problem (2.1.16)–(2.1.21) or (2.1.16)–(2.1.20), (2.1.22) satisfies the estimate*

$$e^{\gamma t}\left(\|u(t)\|^2 + \|v(t) - \bar{v}\|_{H^1}^2 + \|\vec{w}(t)\|^2 + \|\theta(t) - \bar{\theta}\|^2\right)$$

$$+ \int_0^t e^{\gamma t}\left(\|u_y\|^2 + \|v_y\|^2 + \|\vec{w}_y\|^2 + \|\theta_y\|^2\right)(\tau)d\tau \leq C_1, \quad \forall\ t > 0. \quad (2.2.17)$$

Proof. By (2.1.19) and (2.1.23), we get

$$\left[e + \frac{1}{2}(u^2 + |\vec{w}|^2)\right]_t = (-up + \lambda\rho uu_y + \mu\vec{w} \cdot \vec{w}_y + \kappa\rho\theta_y)_y.$$

That is,

$$e_\eta\eta_t + e_v v_t + uu_t + \vec{w} \cdot \vec{w}_t = (-up + \lambda\rho uu_y + \mu\vec{w} \cdot \vec{w}_y + \kappa\rho\theta_y)_y. \quad (2.2.18)$$

Thus it follows from (2.1.17)–(2.1.18) that

$$e_\eta\eta_t + e_v v_t = -pu_y + \lambda\rho u_y^2 + (\kappa\rho\theta_y)_y. \quad (2.2.19)$$

By (2.1.16) and (2.2.8), we get

$$\eta_t = \frac{\lambda\rho u_y^2}{\theta} + \left(\frac{\kappa\rho\theta_y}{\theta}\right)_y + \kappa\rho\left(\frac{\theta_y}{\theta}\right)^2. \quad (2.2.20)$$

Since $\bar{u}, \bar{\theta}, e(\bar{v}, \bar{\eta})$ are constants, $\bar{u}_t = 0, \bar{\theta}_t = 0, e_t(\bar{v}, \bar{\eta}) = 0$, by (2.2.8), (2.2.9) and (2.2.20), we get

$$V_t = \left(\frac{u^2}{2} + \frac{|\vec{w}|^2}{2} + e\right)_t - \frac{\partial e}{\partial v}(\bar{v}, \bar{\eta})v_t - \frac{\partial e}{\partial \eta}(\bar{v}, \bar{\eta})\eta_t$$

$$= \left(\frac{u^2}{2} + \frac{|\vec{w}|^2}{2} + e\right)_t - \bar{\theta}\eta_t + \bar{p}v_t, \quad (2.2.21)$$

which, together with (2.2.20), gives

$$V_t + \frac{\bar{\theta}}{\theta}\left(\lambda\rho u_y^2 + \frac{\kappa\rho\theta_y^2}{\theta}\right) = \left[\lambda\rho u u_y + \mu\rho\vec{w}\cdot\vec{w}_y + (1-\frac{\bar{\theta}}{\theta})\kappa\rho\theta_y - (p-\bar{p})u\right]_y.$$

(2.2.22)

Differentiating ρ_θ with respect to y, we have

$$\frac{\partial}{\partial y}(\rho_\theta) = -2\rho\rho_y u_y - \rho^2 u_{yy}$$

which implies

$$\left(\frac{\rho_y}{\rho}\right)_t = -\rho_y u_y - \rho u_{yy}.$$

(2.2.23)

Multiplying (2.1.17) by $\lambda\rho_y/\rho$, we have

$$\frac{\lambda\rho_y u_t}{\rho} = \frac{\lambda^2\rho\rho_y u_{yy}}{\rho} + \frac{\lambda^2\rho_y^2 u_y}{\rho} - \frac{\lambda p_\rho\rho_y^2}{\rho} - \frac{\lambda p_\theta\theta_y\rho_y}{\rho}.$$

(2.2.24)

We can also get

$$\lambda(-\rho_y u_y - \rho u_{yy})u = -\lambda(\rho u u_y)_y + \lambda\rho u_y^2.$$

(2.2.25)

Differentiating $\frac{\lambda^2}{2}(\frac{\rho_y}{\rho})^2 + \frac{\lambda\rho_y u}{\rho}$ with respect to t, we derive from (2.2.23),

$$\left[\frac{\lambda^2}{2}(\frac{\rho_y}{\rho})^2 + \frac{\lambda\rho_y u}{\rho}\right]_t = \lambda^2(\frac{\rho_y}{\rho})(-\rho_y u_y - \rho u_{yy}) + \frac{\lambda\rho_y u_t}{\rho} + \lambda(-\rho_y u_y - \rho u_{yy})u.$$

(2.2.26)

By (2.2.25) and (2.2.26), we derive

$$\left[\frac{\lambda^2}{2}(\frac{\rho_y}{\rho})^2 + \frac{\lambda\rho_y u}{\rho}\right]_t + \frac{\lambda p_\rho\rho_y^2}{\rho} + \frac{\lambda p_\theta\theta_y\rho_y}{\rho} - \lambda\rho u_y^2 = -\lambda(\rho u u_y)_y.$$

(2.2.27)

Taking the inner product of $\vec{w}$ in $\mathbb{R}^2$ on both sides of (2.1.18), we get

$$\frac{1}{2}\frac{\partial}{\partial t}|\vec{w}|^2 + \mu\rho|\vec{w}_y|^2 = (\mu\rho\vec{w}\cdot\vec{w}_y)_y.$$

(2.2.28)

Multiplying (2.2.22), (2.2.27), (2.2.28) by $e^{\gamma t}, \beta e^{\gamma t}, e^{\gamma t}$ respectively, and then adding the results up, we deduce that

$$\frac{\partial}{\partial t}G(t) + e^{\gamma t}\left[\frac{\bar{\theta}}{\theta}\left(\lambda\rho u_y^2 + \frac{\kappa\rho\theta_y^2}{\theta}\right) + \beta\left(\frac{\lambda p_\rho\rho_y^2}{\rho} + \frac{\lambda p_\theta\theta_y\rho_y}{\rho} - \lambda\rho u_y^2\right) + \mu\rho|\vec{w}_y|^2\right]$$

$$= \gamma G(t) + e^{\gamma t}[(1-\beta)\lambda\rho u u_y + 2\mu\rho\vec{w}\cdot\vec{w}_y + \kappa\left(1-\frac{\bar{\theta}}{\theta}\right)\rho\theta_y - (p-p(\bar{\rho},\bar{\theta}))u]_y,$$

(2.2.29)

where

$$G(t) = e^{\gamma t}\left[V\left(\frac{1}{\rho}, u, \vec{w}, \eta\right) + \beta\left(\frac{\lambda^2}{2}\left(\frac{\rho_y}{\rho}\right)^2\right) + \frac{\lambda\rho_y u}{\rho}\right].$$

Integrating (2.2.29) over $[0, 1] \times [0, t]$, we get

$$\int_0^1 G(t)\,dy$$

$$+ \int_0^t\int_0^1 e^{\gamma\tau}\left[\frac{\bar{\theta}}{\theta}\left(\lambda\rho u_y^2 + \frac{\kappa\rho\theta_y^2}{\theta}\right) + \beta\left(\frac{\lambda p_\rho\rho_y^2}{\rho} + \frac{\lambda p_\theta\theta_y\rho_y}{\rho} - \lambda\rho u_y^2\right) + \mu\rho|\vec{w}_y|^2\right]dyd\tau$$

$$+ \gamma\int_0^t\int_0^1 e^{\gamma t}\left[(1-\beta)\lambda\rho u u_y + 2\mu\rho\vec{w}\cdot\vec{w}_y + \kappa\left(1 - \frac{\bar{\theta}}{\theta}\right)\rho\theta_y - (p - p(\bar{\rho}, \bar{\theta}))u\right]_y d\tau$$

$$= \int_0^1 G(0)\,dy + \gamma\int_0^t\int_0^1 G(\tau)\,dyd\tau. \qquad (2.2.30)$$

Using Cauchy's and Poincaré's inequalities, we deduce that for small $\beta > 0$ and for any $\gamma > 0$,

$$e^{\gamma t}\left(\|\rho(t) - \bar{\rho}\|^2 + \|u(t)\|^2 + \|\vec{w}(t)\|^2 + \|\eta(t) - \bar{\eta}\|^2 + \|\rho_y(t)\|^2\right) \qquad (2.2.31)$$

$$+ \int_0^t e^{\gamma\tau}\left(\|u_y\|^2 + \|\vec{w}_y\|^2 + \|\theta_y\|^2 + \|v_y\|^2\right)(\tau)\,d\tau$$

$$\leq C_1 + C_1\gamma\int_0^t e^{\gamma\tau}\left(\|\rho - \bar{\rho}\|^2 + \|u\|^2 + \|\vec{w}\|^2 + \|\eta - \bar{\eta}\|^2 + \|\rho_y\|^2\right)(\tau)\,d\tau.$$

By Lemma 2.2.1, boundary conditions (2.1.21)–(2.1.22) together with the Poincaré inequality, we get

$$\|v(t) - \bar{v}\| \leq C_1\|v_y(t)\|, \qquad \|\vec{w}(t)\| \leq C_1\|\vec{w}_y(t)\|,$$
$$\|u(t)\| \leq C_1\|u_y(t)\|, \quad \|\theta(t) - \bar{\theta}\| \leq C_1\|\theta_y(t)\|. \qquad (2.2.32)$$

Using the mean value theorem, we infer that

$$\eta(v, \theta) - \bar{\eta} = \eta_v(\tilde{v}, \tilde{\theta})(v - \bar{v}) + \eta_\theta(\tilde{v}, \tilde{\theta})(\theta - \bar{\theta}).$$

Hence,

$$C_1^{-1}(\|v - \bar{v}\|^2 + \|\theta - \bar{\theta}\|^2) \leq \|\eta - \bar{\eta}\|^2 \leq C_1(\|v - \bar{v}\|^2 + \|\theta - \bar{\theta}\|^2), \qquad (2.2.33)$$
$$C_1^{-1}\|v - \bar{v}\|^2 \leq \|\rho - \bar{\rho}\|^2 \leq C_1\|v - \bar{v}\|^2. \qquad (2.2.34)$$

By (2.2.31)–(2.2.34), we get

$$e^{\gamma t}\left(\|\rho(t) - \bar\rho\|^2 + \|u(t)\|^2 + \|\vec w(t)\|^2 + \|\eta(t) - \bar\eta\|^2 + \|\rho_y(t)\|^2\right)$$

$$+ \int_0^t e^{\gamma\tau}\left(\|u_y^2\| + \|\vec w_y\|^2 + \|\theta_y\|^2 + \|v_y\|^2\right)(\tau)\,d\tau$$

$$\le C_1 + C_1\gamma\int_0^t e^{\gamma\tau}\left(\|u_y^2\| + \|\vec w_y\|^2 + \|\theta_y\|^2 + \|v_y\|^2\right)(\tau)\,d\tau$$

which gives that there exists a constant $\gamma_1' = \gamma_1'(C_1) > 0$, such that for any fixed $\gamma \in (0, \gamma']$,

$$e^{\gamma t}\left(\|u(t)\|^2 + \|v(t) - \bar v\|_{H^1}^2 + \|\vec w(t)\|^2 + \|\theta(t) - \bar\theta\|^2\right)$$

$$+ \int_0^t e^{\gamma\tau}\left(\|u_y\|^2 + \|v_y\|^2 + \|\vec w_y\|^2 + \|\theta_y\|^2\right)(\tau)\,d\tau \le C_1.$$

The proof is complete. $\qquad\qquad\qquad\qquad\qquad\qquad\qquad\qquad\qquad\qquad\square$

Lemma 2.2.4. *There exist constants $C_1 > 0$ and $\gamma_1 = \gamma_1(C_1) \le \gamma_1'$ such that for any fixed $\gamma \in (0, \gamma_1]$, the global solution $(u(t), v(t), \vec w(t), \theta(t)) \in H_+^1$ to problem (2.1.16)–(2.1.21) or (2.1.16)–(2.1.20), (2.1.22) satisfies the following estimates:*

$$e^{\gamma t}\left(\|u_y(t)\|^2 + \|\vec w_y(t)\|^2 + \|\theta_y(t)\|^2\right) + \int_0^t e^{\gamma\tau}\left(\|u_{yy}\|^2 + \|\vec w_{yy}\|^2\right.$$

$$\left. + \|\theta_{yy}\|^2 + \|u_t\|^2 + \|\theta_t\|^2 + \|\vec w_t\|^2\right)(\tau)\,d\tau \le C_1, \quad \forall\ t > 0. \qquad (2.2.35)$$

Proof. By Lemma 2.2.1's boundary conditions together with the Poincaré inequality, we get

$$\|\vec w_y(t)\| \le C_1\|\vec w_{yy}(t)\|, \quad \|u_y(t)\| \le C_1\|u_{yy}(t)\|, \quad \|\theta_y(t)\| \le C_1\|\theta_{yy}(t)\|. \qquad (2.2.36)$$

Multiplying (2.1.17) by $-u_{yy}$ and integrating the resulting equality over $[0, 1]$, we have

$$\frac{1}{2}\frac{d}{dt}\|u_y(t)\|^2 + \lambda\int_0^1 \frac{u_{yy}^2}{v}\,dy = \int_0^1 \left(\frac{\lambda u_y v_y}{v^2} + p_\theta\theta_y + p_v v_y\right)u_{yy}\,dy. \qquad (2.2.37)$$

Using the interpolation inequality, we have that for any $\varepsilon > 0$,

$$\int_0^1 \frac{\lambda u_y v_y}{v^2}u_{yy}\,dy \le C_1\|v_y\|\|u_y\|_{L^\infty}\|u_{yy}\|$$

$$\le C_1\|v_y\|\|u_y\|^{\frac{1}{2}}\|u_{yy}\|^{\frac{3}{2}}$$

$$\le C_1(\|v_y\|^2 + \|u_y\|^2) + \varepsilon\|u_{yy}\|^2. \qquad (2.2.38)$$

By Young's inequality and Cauchy's inequality, we have

$$\frac{d}{dt}\|u_y(t)\|^2 + C_1^{-1}\|u_{yy}(t)\|^2 \le C_1(\|\theta_y(t)\|^2 + \|v_y(t)\|^2 + \|u_y(t)\|^2). \qquad (2.2.39)$$

Multiplying (2.2.39) by $e^{\gamma t}$ and integrating the resulting equality over $[0, t]$, implies that there exists a $\gamma_1 = \gamma_1(C_1) \leq \gamma_1'$, such that for any fixed $\gamma \in (0, \gamma_1]$,

$$e^{\gamma t}\|u_y(t)\|^2 + \int_0^t e^{\gamma \tau}\|u_{yy}(\tau)\|^2\, d\tau \tag{2.2.40}$$

$$\leq C_1 \int_0^t e^{\gamma \tau}\left(\|\theta_y\|^2 + \|v_y\|^2 + \|u_y\|^2\right)(\tau)\, d\tau + \gamma \int_0^t e^{\gamma \tau}\|u_y(\tau)\|^2\, d\tau \leq C_1.$$

By (2.1.17),

$$\|u_t(t)\| \leq C_1(\|u_{yy}(t)\| + \|\theta_y(t)\| + \|v_y(t)\| + \|u_y(t)\|),$$

which, along with (2.2.40), gives

$$\int_0^t e^{\gamma \tau}\|u_t(\tau)\|^2\, d\tau \leq C_1, \quad \forall t > 0. \tag{2.2.41}$$

Similarly to (2.2.40)–(2.2.41), we have for $\gamma \in (0, \gamma_1]$ that

$$e^{\gamma t}\|\vec{w}_y(t)\|^2 + \int_0^t e^{\gamma \tau}(\|\vec{w}_{yy}\|^2 + \|\vec{w}_t(\tau)\|^2)(\tau)\, d\tau \leq C_1, \forall t > 0. \tag{2.2.42}$$

Multiplying (2.1.19) by $-e_\theta \theta_{yy}$ and integrating the resulting equality over $[0,1]$, we get

$$\frac{1}{2}\frac{d}{dt}\|\theta_y(t)\|^2 + \int_0^1 \frac{\kappa \theta_{yy}^2}{e_\theta v}\, dy$$

$$= \int_0^1 \left(\frac{e_v v_t}{e_\theta}\theta_{yy} - \frac{\lambda u_y^2 + \mu \vec{w}_y^2}{e_\theta v}\theta_{yy} + \frac{p u_y \theta_{yy}}{e_\theta} + \frac{\kappa \theta_y v_y}{e_\theta v^2}\theta_{yy}\right)\, dy. \tag{2.2.43}$$

Using the interpolation inequality, we have that for any $\varepsilon > 0$,

$$\int_0^1 \frac{\kappa \theta_y v_y}{e_\theta v^2}\theta_{yy}\, dy \leq C_1\|v_y(t)\|\|\theta_y(t)\|_{L^\infty}\|\theta_{yy}(t)\|$$
$$\leq C_1\|v_y(t)\|\|\theta_y(t)\|^{\frac{1}{2}}\|\theta_{yy}(t)\|^{\frac{3}{2}}$$
$$\leq C_1(\|v_y(t)\|^2 + \|\theta_y(t)\|^2) + \varepsilon\|\theta_{yy}(t)\|^2, \tag{2.2.44}$$

$$\int_0^1 \frac{\lambda u_y^2}{e_\theta v}\theta_{yy}\, dy \leq C_1\|u_y(t)\|\|u_y(t)\|_{L^\infty}\|\theta_{yy}(t)\|$$
$$\leq C_1\|u_y(t)\|\|u_y(t)\|^{\frac{1}{2}}\|u_{yy}(t)\|^{\frac{1}{2}}\|\theta_{yy}(t)\|$$
$$\leq C_1(\|u_y(t)\|^2 + \|u_{yy}(t)\|^2) + \varepsilon\|\theta_{yy}(t)\|^2, \tag{2.2.45}$$

$$\int_0^1 \frac{\mu|\vec{w}_y|^2}{e_\theta v}\theta_{yy}\, dy \leq C_1\|\vec{w}_y(t)\|\|\vec{w}_y(t)\|_{L^\infty}\|\theta_{yy}(t)\|$$
$$\leq C_1\|\vec{w}_y(t)\|\|\vec{w}_y(t)\|^{\frac{1}{2}}\|\vec{w}_{yy}(t)\|^{\frac{1}{2}}\|\theta_{yy}(t)\|$$
$$\leq C_1(\|\vec{w}_y(t)\|^2 + \|\vec{w}_{yy}(t)\|^2) + \varepsilon\|\theta_{yy}(t)\|^2. \tag{2.2.46}$$

Inserting (2.2.44)–(2.2.46) into (2.2.43), and using Lemma 2.2.1, we deduce that

$$\frac{d}{dt}\|\theta_y(t)\|^2 + C_1^{-1}\|\theta_{yy}(t)\|^2 \le C_1(\|u_y(t)\|^2 + \|\theta_y(t)\|^2 + \|v_y(t)\|^2 + \|\vec{w}_y(t)\|^2).$$
(2.2.47)

Multiplying (2.2.47) by $e^{\gamma t}$, and integrating the resulting equation over $[0,t]$, we have that there exists a constant $\gamma_1 = \gamma_1(C_1) \le \gamma_1'$, such that for any fixed $\gamma \in (0, \gamma_1]$,

$$e^{\gamma t}\|\theta_y(t)\|^2 + \int_0^t e^{\gamma\tau}\|\theta_{yy}(\tau)\|^2 d\tau \tag{2.2.48}$$

$$\le C_1 \int_0^t e^{\gamma\tau}(\|u_y\|^2 + \|\vec{w}_y\|^2 + \|v_y\|^2)(\tau)d\tau + \gamma \int_0^t e^{\gamma\tau}\|\theta_y(\tau)\|^2 d\tau \le C_1$$

which with

$$\|\theta_t(t)\| \le C_2(\|\theta_{yy}(t)\| + \|\theta_y(t)\| + \|v_y(t)\| + \|u_y(t)\| + \|\vec{w}_y(t)\|)$$

yields

$$\int_0^t e^{\gamma\tau}\|\theta_t(\tau)\|^2 \, d\tau \le C_1. \tag{2.2.49}$$

The proof is complete.   □

2.3   Proof of Theorem 2.1.2

In this section we shall complete the proof of Theorem 2.1.2 and take that the assumptions in Theorem 2.1.2 to be valid. We begin with the following lemma.

Lemma 2.3.1. *Under the assumptions of Theorem 2.1.2, for any* $(v_0, u_0, \vec{w}_0, \theta_0) \in H_+^2$, *the following estimates hold for any* $t > 0$,

$$\|u_t(t)\|^2 + \|\vec{w}_t(t)\|^2 + \|\theta_t(t)\|^2$$

$$+ \int_0^t \left(\|u_{ty}(t)\|^2 + \|\vec{w}_{ty}(t)\|^2 + \|\theta_{ty}(t)\|^2\right)(\tau)d\tau \le C_2, \tag{2.3.1}$$

$$\|u(t)\|_{H^2}^2 + \|v(t) - \bar{v}\|_{H^2}^2 + \|\vec{w}(t)\|_{H^2}^2 + \|\theta(t) - \bar{\theta}\|_{H^2}^2 \le C_2. \tag{2.3.2}$$

Proof. Differentiating (2.1.17) with respect to t, multiplying the result by u_t and integrating over [0,1], we infer that

$$\frac{1}{2}\frac{d}{dt}\|u_t(t)\|^2 = -\int_0^1 \lambda\frac{u_{ty}^2}{v}dy + \int_0^1 \lambda\frac{u_y^2 u_{ty}}{v^2}dy + \int_0^1 p_\theta\theta_t u_{ty}dy + \int_0^1 p_v v_t u_{ty}dy.$$
(2.3.3)

Using Cauchy's inequality, Young's inequality and the embedding theorem, we conclude that for any $\varepsilon > 0$,

$$\frac{1}{2}\frac{d}{dt}\|u_t(t)\|^2 + C_1^{-1}\|u_{ty}(t)\|^2 \leq \varepsilon\|u_{ty}(t)\|^2 + C_1(\|u_y(t)\|_{L^4}^4 + \|\theta_t(t)\|^2 + \|u_y(t)\|^2)$$
$$\leq \varepsilon\|u_{ty}(t)\|^2 + C_2(\|u_{yy}(t)\|^2 + \|\theta_t(t)\|^2 + \|u_y(t)\|^2). \tag{2.3.4}$$

By equations (2.1.16)–(2.1.19), we get

$$\|u_{yy}(t)\| \leq C_2(\|u_t(t)\| + \|\theta_y(t)\| + \|v_y(t)\| + \|u_y(t)\|), \|\vec{w}_{yy}(t)\| \leq C_2\|\vec{w}_t(t)\|, \tag{2.3.5}$$

$$\|\theta_{yy}(t)\| \leq C_2(\|\theta_t(t)\| + \|\theta_y(t)\| + \|v_y(t)\| + \|u_y(t)\| + \|\vec{w}_y(t)\|) \tag{2.3.6}$$

or

$$\|u_t\| \leq C_2(\|u_{yy}(t)\| + \|\theta_y(t)\| + \|v_y(t)\| + \|u_y(t)\|), \|\vec{w}_t(t)\| \leq C_2\|\vec{w}_{yy}(t)\|, \tag{2.3.7}$$

$$\|\theta_t(t)\| \leq C_2(\|\theta_{yy}(t)\| + \|\theta_y(t)\| + \|v_y(t)\| + \|u_y(t)\| + \|\vec{w}_y(t)\|). \tag{2.3.8}$$

Since $(v_0, u_0, \vec{w}_0, \theta_0) \in H_+^2$, we can infer from (2.3.7)–(2.3.8) that

$$\|u_t(y,0)\| \leq C_2, \quad \|\vec{w}_t(y,0)\| \leq C_2, \quad \|\theta_t(y,0)\| \leq C_2. \tag{2.3.9}$$

Integrating (2.3.4) on $[0,t]$, using Lemma 2.2.1 and (2.3.7)–(2.3.9), we get

$$\|u_t(t)\|^2 + \int_0^t \|u_{ty}(\tau)\|^2 d\tau \leq C_2, \quad \forall t > 0, \tag{2.3.10}$$

which, together with (2.3.5), implies

$$\|u_{yy}(t)\| \leq C_2, \forall t > 0. \tag{2.3.11}$$

Differentiating (2.1.18) with respect to t, multiplying the resulting equation by $\vec{w}_t$ and integrating the resulting equality over $[0,1]$, we infer that

$$\frac{1}{2}\frac{d}{dt}\|\vec{w}_t(t)\|^2 = -\int_0^1 \mu\frac{|\vec{w}_{ty}|^2}{v}dy + \int_0^1 \mu\frac{\vec{w}_y \cdot \vec{w}_{ty}v_t}{v^2}dy,$$

which, by using Cauchy's inequality, Young's inequality and the embedding theorem, gives that for $\varepsilon > 0$ small enough,

$$\frac{1}{2}\frac{d}{dt}\|\vec{w}_t(t)\|^2 + C_1^{-1}\|\vec{w}_{ty}(t)\|^2 \leq \varepsilon\|\vec{w}_{ty}(t)\|^2 + C_1\|v_t\vec{w}_y\|^2$$
$$\leq \varepsilon\|\vec{w}_{ty}(t)\|^2 + C_1\|v_t(t)\|_{L^\infty}^2\|\vec{w}_y(t)\|^2. \tag{2.3.12}$$

Thus integrating (2.3.12) over $[0,t]$ and using Lemma 2.2.1, we get

$$\|\vec{w}_t(t)\|^2 + \int_0^t \|\vec{w}_{ty}(\tau)\|^2 d\tau \leq C_2\int_0^t \|\vec{w}_y(\tau)\|^2 d\tau + C_1\|\vec{w}_t(y,0)\|^2 \leq C_2, \tag{2.3.13}$$

which, along with (2.3.5), gives

$$\|\vec{w}_{yy}(t)\| \le C_2, \forall t > 0. \tag{2.3.14}$$

By (2.1.17)–(2.1.19) and (2.1.23), we get

$$e_\theta \theta_t + e_v v_t = -p u_y + \frac{\lambda u_y^2}{v} + \frac{\mu |\vec{w}_y|^2}{v} + \left(\frac{\kappa \theta_y}{v}\right)_y. \tag{2.3.15}$$

Differentiating (2.3.15) with respect to t, multiplying the result θ_t, and finally integrating the resultant over [0,1], we get

$$\frac{1}{2}\frac{d}{dt}\|\sqrt{e_\theta}\theta_t\|^2 + \int_0^1 \frac{\kappa \theta_{ty}^2}{v} dy = \int_0^1 \left(2\lambda \frac{u_y u_{yt}}{v} - \lambda \frac{u_y^2 v_t}{v^2} + 2\mu \frac{\vec{w}_y \cdot \vec{w}_{yt}}{v} - \mu \frac{|\vec{w}_y|^2 v_t}{v^2}\right.$$
$$- u_{yt}p - u_y p_\theta \theta_t - u_y p_v v_t - e_v v_{tt} - e_{vv} v_t^2$$
$$\left. - \frac{1}{2} e_{\theta\theta} \theta_t^2 - \frac{3}{2} e_{\theta v} v_t \theta_t + \frac{\kappa \theta_{ty} \theta_y v_t}{v^2}\right)\theta_t dy. \tag{2.3.16}$$

Integrating (2.3.16) over [0,t], using the Cauchy inequality, the Young inequality and the embedding theorem, we conclude that for any $\varepsilon > 0$,

$$\frac{1}{2}\frac{d}{dt}\|\sqrt{e_\theta}\theta_t\|^2 + C_1^{-1}\int_0^t \|\theta_{ty}(\tau)\|^2 d\tau$$
$$\le \varepsilon \int_0^t \|\theta_{ty}(\tau)\|^2 d\tau + C_1 \int_0^t \left(\|\theta_y\|^2 + \|\theta_{yy}\|^2 + \|\theta_t\|^2\right.$$
$$\left. + \|u_y\|^2 + \|u_{ty}\|^2 + \|\vec{w}_y\|^2 + \|\vec{w}_{ty}\|^2 + \|\theta_t(\tau)\|_{L^3}^3\right)d\tau. \tag{2.3.17}$$

By the Nirenberg interpolation inequality, we can get for any $\varepsilon > 0$,

$$\int_0^t \|\theta_t\|_{L^3}^3(\tau)d\tau \le C_1 \int_0^t \left(\|\theta_t\|^{\frac{5}{2}}\|\theta_{ty}\|^{\frac{1}{2}} + \|\theta_t\|^3\right)(\tau)d\tau$$
$$\le C_1 \sup_{0\le\tau\le t} \|\theta_t(\tau)\|^{\frac{4}{3}}\int_0^t \left(\|\theta_t\|^2 + \|\theta_t\|^{\frac{5}{3}}\right)(\tau)d\tau + \varepsilon \int_0^t \|\theta_{ty}(\tau)\|^2 d\tau$$
$$\le C_1 \sup_{0\le\tau\le t} \|\theta_t(\tau)\|^{\frac{4}{3}} + \varepsilon \int_0^t \|\theta_{ty}(\tau)\|^2 d\tau$$
$$\le \frac{1}{2} \sup_{0\le\tau\le t} \|\theta_t(\tau)\|^2 + \varepsilon \int_0^t \|\theta_{ty}(\tau)\|^2 d\tau + C_2. \tag{2.3.18}$$

Hence, inserting (2.3.18) into (2.3.17), and taking $\varepsilon > 0$ small enough, we get

$$\|\theta_t(t)\|^2 + \int_0^t \|\theta_{ty}(\tau)\|^2 d\tau \le C_2, \quad \forall t > 0. \tag{2.3.19}$$

By (2.3.6) and (2.3.19), we conclude that

$$\|\theta_{yy}(t)\| \le C_2, \quad \forall t > 0.$$

The proof is complete. □

Lemma 2.3.2. *Under the assumptions of Theorem 2.1.2, for any* $(u_0, v_0, \vec{w}_0, \theta_0) \in H_+^2$, *the following estimate holds for any* $t > 0$:

$$\|v(t)\|_{H^2}^2 + \int_0^t \|v(\tau)\|_{H^2}^2 d\tau \le C_2.$$

Proof. Differentiating (2.1.17) with respect to y, by equation (2.1.16),

$$v_{tyy} = u_{yyy},$$

we get

$$\lambda \frac{\partial}{\partial t} \left(\frac{v_{yy}}{v} \right) - p_v v_{yy} = u_{ty} + p_{vv} v_y^2 + 2 p_{v\theta} v_y \theta_y + p_\theta \theta_{yy} + p_{\theta\theta} \theta_y^2 + 2\lambda \frac{u_{yy} v_y}{v^2} - 2\lambda \frac{u_y v_y^2}{v^3} \tag{2.3.20}$$

where

$$\lambda \frac{\partial}{\partial t} \left(\frac{v_{yy}}{v} \right) = \lambda \frac{v_{yyt}}{v} - \lambda \frac{v_{yy} v_t}{v^2}.$$

Multiplying (2.3.20) by $\dfrac{v_{yy}}{v}$, we get

$$\frac{\lambda}{2} \frac{d}{dt} \left\| \frac{v_{yy}}{v} \right\|^2 + C_1^{-1} \left\| \frac{v_{yy}}{v} \right\|^2 \le \frac{C_1^{-1}}{4} \left\| \frac{v_{yy}}{v} \right\|^2 + C_2 \Big(\|u_{yy} v_y\|^2 + \|u_y v_y^2\|^2 + \|\theta_y v_y\|^2$$
$$+ \|v_y\|_{L^4}^4 + \|\theta_y\|_{L^4}^4 + \|\theta_{yy}\|^2 + \|u_{ty}\|^2 \Big). \tag{2.3.21}$$

Integrating (2.3.21) over $[0, t]$ gives

$$\|v_{yy}(t)\|^2 + \int_0^t \|v_{yy}(\tau)\|^2 d\tau \le C_2 \int_0^t \Big(\|u_{yy}\|^2 + \|u_y\|_{L^\infty} \|v_y\|_{L^4}^4 + \|\theta_y\|_{L^\infty} \|v_y\|^2$$
$$+ \|v_y\|_{L^4}^4 + \|\theta_y\|_{L^4}^4 + \|\theta_{yy}\|^2 + \|u_{ty}\|^2 \Big)(\tau) d\tau.$$

Using the Young inequality, the Poincaré inequality, the Nirenberg interpolation inequality, Lemmas 2.2.1 and 2.3.1, we infer that

$$\|v_{yy}(t)\|^2 + \int_0^t \|v_{yy}(\tau)\|^2 d\tau \le C_2 \int_0^t \Big(\|v_y\|^2 + \|u_{ty}\|^2 + \|u_{yy}\|^2 + \|\theta_{yy}\|^2 \Big)(\tau) d\tau$$
$$\le C_2.$$

The proof is complete. □

Lemma 2.3.3. *Under the assumptions of Theorem 2.1.2, for any* $(v_0, u_0, \vec{w}_0, \theta_0) \in$ H_+^2, *the following estimate holds for any* $t > 0$:

$$\int_0^t \left(\|u(\tau)\|_{H^3}^2 + \|\vec{w}(\tau)\|_{H^3}^2 + \|\theta(\tau) - \bar{\theta}\|_{H^3}^2 \right) d\tau \leq C_2. \tag{2.3.22}$$

Proof. Differentiating (2.1.17) with respect to y, we infer that

$$\frac{\lambda u_{yyy}}{v} = u_{ty} + p_\theta \theta_{yy} + p_v v_{yy} + 2p_{\theta v} \theta_y v_y + p_{\theta\theta} \theta_y^2 + p_{vv} v_y^2$$
$$+ \frac{2\lambda u_{yy} v_y}{v^2} + \frac{\lambda u_y v_{yy}}{v^2} - \frac{2\lambda u_y v_y^2}{v^3}. \tag{2.3.23}$$

Multiplying (2.3.23) by u_{yyy}, integrating the resultant over $[0,1]$, using Cauchy's inequality and Young's inequality, we get for any $\varepsilon > 0$,

$$\|u_{yyy}(t)\|^2 \leq C_2 \Big(\|u_{ty}(t)\|^2 + \|\theta_{yy}(t)\|^2 + \|v_{yy}(t)\|^2 + \|\theta_y v_y\|^2 + \|\theta_y(t)\|_{L^4}^4$$
$$+ \|v_y(t)\|_{L^4}^4 + \|u_{yy} v_y\|^2 + \|v_{yy} u_y\|^2 + \|u_y v_y^2\|^2 \Big) + \varepsilon \|u_{yyy}(t)\|^2. \tag{2.3.24}$$

Integrating (2.3.24) over $[0, t]$, by Lemma 2.2.1 and Lemma 2.3.1, we get

$$\int_0^t \|u_{yyy}(\tau)\|^2 d\tau \leq C_2 \int_0^t \left(\|u_{ty}\|^2 + \|\theta_{yy}\|^2 + \|v_{yy}\|^2 + \|u_{yy}\|^2 \right)(\tau) d\tau \leq C_2. \tag{2.3.25}$$

Differentiating (2.1.18) with respect to y, we infer that

$$\frac{\mu \vec{w}_{yyy}}{v} = \vec{w}_{ty} + \frac{2\mu \vec{w}_{yy} v_y}{v^2} + \frac{\mu \vec{w}_y v_{yy}}{v^2} - \frac{2\mu \vec{w}_y v_y^2}{v^3}. \tag{2.3.26}$$

Multiplying (2.3.26) by $\vec{w}_{yyy}$, integrating the resulting equality over $[0, 1]$, using Cauchy's inequality and Young's inequality, we get that for any small $\varepsilon > 0$,

$$\|\vec{w}_{yyy}(t)\|^2 \leq C_2 \left(\|\vec{w}_{ty}(t)\|^2 + \|v_{yy}(t)\|^2 + \|\vec{w}_{yy}(t)\|^2 \right). \tag{2.3.27}$$

Integrating (2.3.27) over $[0, t]$, by Lemma 2.2.1 and Lemma 2.3.1, we arrive at

$$\int_0^t \|\vec{w}_{yyy}(\tau)\|^2 d\tau \leq C_2 \int_0^t (\|\vec{w}_{ty}\|^2 + \|v_{yy}\|^2 + \|\vec{w}_{yy}\|^2)(\tau) d\tau \leq C_2. \tag{2.3.28}$$

Differentiating (2.1.19) with respect to y, we obtain

$$\frac{\kappa \theta_{yyy}}{v} = e_\theta \theta_{ty} + e_v v_{ty} + e_{\theta v} \theta_t v_y + e_{\theta v} \theta_y v_t + e_{\theta\theta} \theta_y \theta_t + e_{vv} v_y v_t$$
$$- \frac{2\lambda u_{yy} u_y}{v^2} + \frac{\lambda v_y u_y^2}{v^2} - \frac{2\mu \vec{w}_{yy} \cdot \vec{w}_y}{v^2} + \frac{\mu v_y |\vec{w}_y|^2}{v^2} + \frac{2\kappa \theta_{yy} v_y}{v^2}$$
$$+ \frac{\kappa \theta_y v_{yy}}{v^2} - \frac{2\kappa \theta_y v_y^2}{v^3} + u_{yy} p + p_\theta \theta_y u_y + p_v u_y v_y. \tag{2.3.29}$$

Multiplying (2.3.29) by θ_{yyy}, integrating the resultant over $[0,1]$ and using Cauchy's inequality, we deduce that for any small $\varepsilon > 0$,

$$\|\theta_{yyy}(t)\|^2 \le C_2 \Big(\|u_{ty}(t)\|^2 + \|\theta_{ty}(t)\|^2 + \|v_{yy}(t)\|^2 + \|\theta_y v_y\|^2 + \|\theta_y(t)\|_{L^4}^4$$
$$+ \|v_y(t)\|_{L^4}^4 + \|u_{yy} v_y\|^2 + \|v_{yy} u_y\|^2 + \|u_y v_y^2\|^2 \Big) + \varepsilon \|\theta_{yyy}(t)\|^2,$$

i.e.,

$$\|\theta_{yyy}(t)\|^2 \le C_2 \Big(\|u_{ty}(t)\|^2 + \|\theta_{ty}(t)\|^2 + \|v_{yy}(t)\|^2 + \|v_{yy}(t)\|^2$$
$$+ \|u_y(t)\|^2 + \|v_y(t)\|^2 + \|\theta_y(t)\|^2 \Big). \tag{2.3.30}$$

Integrating (2.3.30) over $[0,t]$, and using Lemma 2.2.1 and Lemma 2.3.1, we get

$$\int_0^t \|\theta_{yyy}(\tau)\|^2 d\tau \le C_2 \int_0^t \Big(\|u_{ty}\|^2 + \|\theta_{ty}\|^2 + \|v_{yy}\|^2 + \|u_{yy}\|^2 + \|\theta_y\|^2$$
$$+ \|u_y\|^2 + \|v_y\|^2 \Big)(\tau) d\tau \le C_2.$$

The proof is complete. $\square$

Lemma 2.3.4. *There exist constants $C_2 > 0$ and $\gamma_2' = \gamma_2'(C_2) > 0$ such that for any fixed $\gamma \in (0, \gamma_2']$, the global solution $(v(t), u(t), \vec{w}(t), \theta(t)) \in H_+^2$ to problem (2.1.16)–(2.1.21) or (2.1.16)–(2.1.20), (2.1.22) satisfies that the following estimates:*

$$e^{\gamma t} \Big(\|v(t) - \bar{v}\|_{H^2}^2 + \|u(t)\|_{H^2}^2 + \|\vec{w}(t)\|_{H^2}^2 + \|\theta(t) - \bar{\theta}\|_{H^2}^2 \Big) \tag{2.3.31}$$
$$+ \int_0^t e^{\gamma \tau} \Big(\|v - \bar{v}\|_{H^2}^2 + \|u\|_{H^3}^2 + \|\vec{w}\|_{H^3}^2 + \|\theta\|_{H^3}^2 \Big)(\tau) d\tau \le C_2, \quad \forall\ t > 0,$$

$$e^{\gamma t} \Big(\|u_t(t)\|^2 + \|\vec{w}_t(t)\|^2 + \|\theta_t(t)\|^2 \Big) \tag{2.3.32}$$
$$+ \int_0^t e^{\gamma \tau} \Big(\|u_{ty}\|^2 + \|\vec{w}_{ty}\|^2 + \|\theta_{ty}\|^2 \Big)(\tau) d\tau \le C_2, \quad \forall\ t > 0.$$

Proof. Multiplying (2.3.4) by $e^{\gamma t}$, and integrating the resulting equality over $[0,t]$, we have

$$e^{\gamma t} \|u_t(t)\|^2 + \int_0^t e^{\gamma \tau} \|u_{ty}(\tau)\|^2 d\tau$$
$$\le C_2 \int_0^t e^{\gamma \tau} (\|u_{yy}\|^2 + \|\theta_t\|^2 + \|u_y\|^2)(\tau) d\tau + \gamma \int_0^t e^{\gamma \tau} \|u_t(\tau)\|^2 d\tau. \tag{2.3.33}$$

By Lemma 2.2.4 and (2.3.33), we can infer that there exists a constant $\gamma_2' = \gamma_2'(C_1) > 0$ such that for any fixed $\gamma \in (0, \gamma_2']$,

$$e^{\gamma t} \|u_t(t)\|^2 + \int_0^t e^{\gamma \tau} \|u_{ty}(\tau)\|^2 d\tau \le C_2. \tag{2.3.34}$$

Multiplying (2.3.24) by $e^{\gamma t}$, and integrating the resulting equality over $[0,t]$, we get

$$\int_0^t e^{\gamma \tau} \|u_{yyy}(\tau)\|^2 d\tau \le C_2. \tag{2.3.35}$$

Multiplying (2.3.12) by $e^{\gamma t}$, and integrating the result over $[0,t]$, we have

$$e^{\gamma t}\|\vec{w}_t(t)\|^2 + \int_0^t e^{\gamma \tau}\|\vec{w}_{ty}(\tau)\|^2 d\tau \le C_2 \int_0^t e^{\gamma \tau}\|\vec{w}_y(\tau)\|^2 d\tau + \gamma \int_0^t e^{\gamma \tau}\|\vec{w}_t(\tau)\|^2 d\tau. \tag{2.3.36}$$

By Lemma 2.2.4 and (2.3.36), we can derive that there exists a constant $\gamma_2' = \gamma_2'(C_1) > 0$ such that for any fixed $\gamma \in (0, \gamma_2']$,

$$e^{\gamma t}\|\vec{w}_t(t)\|^2 + \int_0^t e^{\gamma \tau}\|\vec{w}_{ty}(\tau)\|^2 d\tau \le C_2. \tag{2.3.37}$$

Multiplying (2.3.28) by $e^{\gamma t}$, and integrating the resultant over $[0,t]$, we get

$$\int_0^t e^{\gamma \tau}\|\vec{w}_{yyy}\tau)\|^2 (d\tau \le C_2. \tag{2.3.38}$$

Multiplying (2.3.17) by $e^{\gamma t}$, and integrating the resultant over $[0,t]$, we have

$$e^{\gamma t}\|\theta_t(t)\|^2 + \int_0^t e^{\gamma \tau}\|\theta_{ty}(\tau)\|^2 \, d\tau$$

$$\le C_2 \int_0^t e^{\gamma \tau}\Big(\|\theta_y\|^2 + \|\theta_{yy}\|^2 + \|\theta_t\|^2 + \|u_y\|^2 + \|u_{ty}\|^2$$

$$+ \|\vec{w}_y\|^2 + \|\vec{w}_{ty}\|^2 + \|\theta_t\|_{L^3}^3\Big)(\tau) \, d\tau. \tag{2.3.39}$$

By Lemma 2.2.4 and (2.3.37), we get that there exists a constant $\gamma_2' = \gamma_2'(C_1) > 0$ such that for any fixed $\gamma \in (0, \gamma_2']$,

$$e^{\gamma t}\|\theta_t(t)\|^2 + \int_0^t e^{\gamma \tau}\|\theta_{ty}(\tau)\|^2 d\tau \le C_2. \tag{2.3.40}$$

Multiplying (2.3.30) by $e^{\gamma t}$, and integrating the resultant over $[0,t]$, we get

$$\int_0^t e^{\gamma \tau}\|\theta_{yyy}(\tau)\|^2 d\tau \le C_2, \quad \forall t > 0. \tag{2.3.41}$$

Multiplying (2.3.21) by $e^{\gamma t}$, and integrating over $[0,t]$, we have

$$e^{\gamma t}\|v_{yy}(t)\|^2 + \int_0^t e^{\gamma \tau}\|v_{yy}(\tau)\|^2 \, d\tau \tag{2.3.42}$$

$$\le C_2 \int_0^t e^{\gamma \tau}(\|v_y\|^2 + \|u_{ty}\|^2 + \|u_{yyy}\|^2 + \|\theta_{yy}\|^2)(\tau)d\tau + \gamma \int_0^t e^{\gamma \tau}\|v_t(\tau)\|^2 d\tau.$$

By Lemma 2.2.3 and (2.3.42), we can deduce that there exists a constant $\gamma_2' = \gamma_2'(C_1) > 0$ such that for any fixed $\gamma \in (0, \gamma_2']$,

$$e^{\gamma t}\|v_{yy}(t)\|^2 + \int_0^t e^{\gamma \tau}\|v_{yy}(\tau)\|^2 d\tau \leq C_2.$$

The proof is complete.   □

2.4   Proof of Theorem 2.1.3

Lemma 2.4.1. *Under the assumptions of Theorem 2.1.3, for any* $(v_0, u_0, \vec{w}_0, \theta_0) \in H_+^4$ *and for any* $\varepsilon \in (0, 1)$ *small enough, we have for any* $t > 0$,

$$\|u_{ty}(y, 0)\| + \|\vec{w}_{ty}(y, 0)\| + \|\theta_{ty}(y, 0)\| \leq C_3, \tag{2.4.1}$$

$$\begin{aligned} \|u_{tt}(y,0)\| + \|\vec{w}_{tt}(y,0)\| + \|\theta_{tt}(y,0)\| + \|u_{tyy}(y,0)\| \\ + \|\vec{w}_{tyy}(y,0)\| + \|\theta_{tyy}(y,0)\| \leq C_4, \end{aligned} \tag{2.4.2}$$

$$\|u_{tt}(t)\|^2 + \int_0^t \|u_{tty}(\tau)\|^2 \, d\tau \leq C_4 + C_4 \int_0^t \|\theta_{tty}(\tau)\|^2 \, d\tau, \tag{2.4.3}$$

$$\|\vec{w}_{tt}(t)\|^2 + \int_0^t \|\vec{w}_{tty}(\tau)\|^2 \, d\tau \leq C_4 + C_4 \int_0^t \|\vec{w}_{tyy}(\tau)\|^2 \, d\tau, \tag{2.4.4}$$

$$\|\theta_{tt}(t)\|^2 + \int_0^t \|\theta_{tty}(\tau)\|^2 \, d\tau \tag{2.4.5}$$

$$\leq C_4 \varepsilon^{-3} + C_2 \varepsilon^{-1} \int_0^t \|\theta_{tyy}(\tau)\|^2 \, d\tau + C_1 \varepsilon \int_0^t \left(\|u_{tty}\|^2 + \|u_{tyy}\|^2 + \|\vec{w}_{tyy}\|^2 \right)(\tau) \, d\tau.$$

Proof. By (2.1.17), we can derive

$$\begin{aligned} \|u_t\| &\leq C_1(\|u_{yy}\| + \|u_y\|_{L^\infty}\|v_y\| + \|v_y\| + \|\theta_y\|) \\ &\leq C_2(\|u_y\|_{H^1} + \|v_y\| + \|\theta_y\|). \end{aligned} \tag{2.4.6}$$

We differentiate (2.1.17) with respect to y, and use Theorems 2.1.1–2.1.2 to derive

$$\begin{aligned} \|u_{ty}\| \leq C_1 \Big(&\|v_y\|_{L^4}^2 + \|v_y\|_{L^\infty}\|\theta_y\| + \|v_{yy}\| + \|u_y\|_{L^\infty}\|v_y\|_{L^4}^2 + \|\theta_y\|_{L^4}^2 \\ &+ \|u_{yyy}\| + \|\theta_{yy}\| + \|v_y\|_{L^\infty}\|u_{yy}\| + \|v_{yy}\|\|u_y\|_{L^\infty}\| \Big). \end{aligned} \tag{2.4.7}$$

Using the Gagliardo-Nirenberg inequality and the Young inequality, we conclude

$$\begin{aligned} \|u_y\|_{L^\infty} &\leq C(\|u_y\|^{\frac{1}{2}}\|u_{yy}\|^{\frac{1}{2}} + \|u_{yy}\|) \leq C(\|u_y\| + \|u_{yy}\|), \\ \|u_y\|_{L^4}^2 &\leq C(\|u_y\|^{\frac{3}{4}}\|u_{yy}\|^{\frac{1}{4}} + \|u_y\|)^2 \leq C(\|u_y\|^{\frac{3}{2}}\|u_{yy}\|^{\frac{1}{2}} + \|u_y\|^2) \\ &\leq C(\|u_y\|^2 + \|u_{yy}\|^2). \end{aligned}$$

Finally, using (2.4.7), we can obtain

$$\|u_{ty}\| \le C_2(\|v_y\|_{H^1} + \|u_y\|_{H^2} + \|\theta_y\|_{H^1}) \tag{2.4.8}$$

or

$$\|u_{yyy}\| \le C_2(\|u_{ty}\| + \|v_y\|_{H^1} + \|u_y\|_{H^1} + \|\theta_y\|_{H^1}). \tag{2.4.9}$$

We differentiate (2.1.17) with respect to y twice to derive

$$\begin{aligned}
\|u_{tyy}\| \le C_1 \Big(&\|v_y\|_{L^6}^3 + \|v_y\|_{L^4}^2\|\theta_y\|_{L^\infty} + \|v_y\|_{L^\infty}\|\|v_{yy}\| + \|\theta_y\|_{L^\infty}\|\|v_{yy}\| \\
&+ \|v_y\|_{L^\infty}\|\theta_y^2\| + \|v_y\|_{L^\infty}\|\|\theta_{yy}\| + \|\theta_y\|_{L^6}^3 + \|v_{yyy}\| + \|\theta_{yyy}\| \\
&+ \|\theta_y\|_{L^\infty}\|\theta_{yy}\| + \|\theta_{yy}\| + \|v_y\|_{L^\infty}\|u_{yyy}\| + \|u_{yy}\|_{L^\infty}\|v_{yy}\| \\
&+ \|u_{yy}\|_{L^\infty}\|v_y\|_{L^4}^2 + \|u_y\|_{L^\infty}\|v_{yyy}\| + \|u_y\|_{L^\infty}\|v_{yy}\|\|v_y\| \\
&+ \|u_{yyyy}\| + \|u_y\|_{L^\infty}\|v_y\|_{L^6}^3 \Big). \tag{2.4.10}
\end{aligned}$$

Using the Gagliardo-Nirenberg and the Young inequality, we conclude

$$\|u_{tyy}\| \le C_2 \left(\|u_y\|_{H^3} + \|v_y\|_{H^2} + \|\theta_y\|_{H^2}\right) \tag{2.4.11}$$

or

$$\|u_{yyyy}\| \le C_2 \left(\|u_{tyy}\| + \|u_y\|_{H^2} + \|v_y\|_{H^2} + \|\theta_y\|_{H^2}\right). \tag{2.4.12}$$

By (2.1.18), we can derive

$$\|\vec{w}_t\| \le C_1(\|\vec{w}_{yy}\| + \|\vec{w}_y\|_{L^\infty}\|v_y\|) \le C_2(\|\vec{w}_y\|_{H^1} + \|v_y\|). \tag{2.4.13}$$

We differentiate (2.1.18) with respect to y, and use Theorems 2.1.1–2.1.2 to get

$$\|\vec{w}_{ty}(t)\| \le C_1 \left(\|\vec{w}_{yyy}\| + \|\vec{w}_{yy}\|\|v_y\|_{L^\infty} + \|\vec{w}_y\|_{L^\infty}\|v_{yy}\| + \|\vec{w}_y\|_{L^\infty}\|v_y^2\|\right).$$

Using the Gagliardo-Nirenberg inequality and the Young inequality, we conclude

$$\|\vec{w}_{ty}(t)\| \le C_2(\|\vec{w}_y\|_{H^2} + \|v_y\|_{H^1}) \tag{2.4.14}$$

or

$$\|\vec{w}_{yyy}\| \le C_2(\|\vec{w}_y\|_{H^1} + \|v_y\|_{H^1} + \|\vec{w}_{ty}\|). \tag{2.4.15}$$

We differentiate (2.1.18) with respect to y twice to derive

$$\begin{aligned}
\|\vec{w}_{tyy}(t)\| &\le C_1 \Big(\|\vec{w}_{yyyy}\| + \|v_y\|_{L^\infty}\|\vec{w}_{yyy}\| + \|v_{yy}\|_{L^\infty}\|\vec{w}_{yy}\| + \|\vec{w}_y\|_{L^\infty}\|\vec{w}_{yyy}\| \\
&\quad + \|\vec{w}_{yy}\|_{L^\infty}\|v_y^2\| + \|\vec{w}_y\|_{L^\infty}\|v_y^3\| + \|\vec{w}_y v_y\|\|v_{yy}\|_{L^\infty} \Big) \\
&\le C_2(\|\vec{w}_y(t)\|_{H^3} + \|v_y(t)\|_{H^2})
\end{aligned}$$

$$\tag{2.4.16}$$

or
$$\|\vec{w}_{yyyy}(t)\| \leq C_2(\|\vec{w}_{tyy}(t)\| + \|\vec{w}_y(t)\|_{H^2} + \|v_y(t)\|_{H^2}). \tag{2.4.17}$$
By (2.1.19), we can derive
$$\|\theta_t(t)\| \leq C_1 \left(\|u_y(t)\| + \|u_y^2(t)\| + \|\vec{w}_y^2(t)\| + \|\theta_{yy}(t)\| + \|\theta_y(t)\|_{L^\infty}\|v_y(t)\| \right)$$
$$\leq C_2(\|u_y(t)\|_{H^1} + \|\vec{w}_y(t)\|_{H^1} + \|\theta_y(t)\|_{H^1}). \tag{2.4.18}$$
Differentiate (2.1.19) with respect to y, and use Theorems 2.1.1–2.1.2 to obtain
$$\|\theta_{ty}(t)\| \leq C_2(\|u_y(t)\|_{H^1} + \|\vec{w}_y(t)\|_{H^1} + \|\theta_y(t)\|_{H^2} + \|v_y(t)\|_{H^1}) \tag{2.4.19}$$
or
$$\|\theta_{yyy}(t)\| \leq C_2 \left(\|\theta_{ty}(t)\| + \|u_y(t)\|_{H^1} + \|\vec{w}_y(t)\|_{H^1} + \|\theta_y(t)\|_{H^1} + \|v_y(t)\|_{H^1} \right). \tag{2.4.20}$$
Differentiate (2.1.18) with respect to y twice to derive
$$\|\theta_{tyy}(t)\| \leq C_2 \left(\|u_y(t)\|_{H^2} + \|\vec{w}_y(t)\|_{H^2} + \|\theta_y(t)\|_{H^3} + \|v_y(t)\|_{H^2} \right) \tag{2.4.21}$$
or
$$\|\theta_{yyyy}(t)\| \leq C_2 \left(\|\theta_{tyy}(t)\| + \|u_y(t)\|_{H^2} + \|\vec{w}_y(t)\|_{H^2} + \|\theta_y(t)\|_{H^2} + \|v_y(t)\|_{H^2} \right). \tag{2.4.22}$$
Differentiate (2.1.17) with respect to t to obtain
$$\|u_{tt}(t)\| \leq C_2 \Big(\|\theta_{ty}(t)\| + \|\theta_t(t)\|_{L^\infty}\|\theta_y(t)\| + \|\theta_y(t)\|_{L^\infty}\|u_y(t)\|$$
$$+ \|v_{ty}(t)\| + \|u_{tyy}(t)\| + \|v_y(t)\|_{L^\infty}\|\theta_t(t)\| + \|u_{ty}(t)\|\|v_y(t)\|_{L^\infty}$$
$$+ \|v_y(t)\|_{L^\infty}\|u_{yy}(t)\| + \|u_y(t)\|_{L^\infty}\|v_y^2(t)\| + \|u_y(t)\|_{L^\infty}\|v_y(t)\| \Big)$$
$$\leq C_2 \Big(\|\theta_y(t)\|_{H^2} + \|v_y(t)\|_{H^2} + \|u_y(t)\|_{H^3} \Big). \tag{2.4.23}$$
We differentiate (2.1.18) with respect to t to deduce
$$\|\vec{w}_{tt}(t)\| \leq C_2 \Big(\|\vec{w}_{tyy}(t)\| + \|u_y(t)\|_{L^\infty}\|\vec{w}_{yy}(t)\| + \|\vec{w}_{ty}(t)\|\|v_y(t)\|_{L^\infty}$$
$$+ \|u_{yy}(t)\|\|\vec{w}_y(t)\|_{L^\infty} + \|\vec{w}_y(t)\|_{L^\infty}\|v_y(t)u_y(t)\| \Big)$$
$$\leq C_2 \left(\|\vec{w}_y(t)\|_{H^3} + \|u_y(t)\|_{H^2} + \|v_y(t)\|_{H^2} \right). \tag{2.4.24}$$
We differentiate (2.1.19) with respect to t to infer
$$\|\theta_{tt}(t)\| \leq C_2 \Big(\|\theta_t\|_{L^4}^2 + \|u_y\|_{L^4}^2 + \|\theta_t\|\|u_y\|_{L^\infty} + \|u_{ty}\| + \|u_{ty}\|\|u_y\|_{L^\infty}$$
$$+ \|\vec{w}_{ty}\|\|\vec{w}_y\|_{L^\infty} + \|u_y\|_{L^\infty}\|\vec{w}_t\|_{L^4}^2 + \|\theta_{tyy}\| + \|u_y\|_{L^\infty}\|\theta_{yy}\|$$
$$+ \|\theta_{yt}\|\|v_y\|_{L^\infty} + \|\theta_y\|_{L^\infty}\|v_{yt}\| + \|u_y\|_{L^\infty}\|v_y\theta_y\| \Big)$$

$$\leq C_2 \Big(\|u_y\|_{H^1} + \|v_y\|_{H^1} + \|\theta_y\|_{H^2} + \|\theta_{tyy}\|$$

$$+ \|u_{ty}\| + \|\theta_{ty}\| + \|\vec{w}_{ty}\| \Big), \tag{2.4.25}$$

$$\leq C_2 \Big(\|u_y(t)\|_{H^2} + \|\vec{w}_y(t)\|_{H^2} + \|\theta_y(t)\|_{H^3} + \|v_y(t)\|_{H^2} \Big). \tag{2.4.26}$$

Thus estimates (2.4.1)–(2.4.2) follow from (2.4.6)–(2.4.25).

Differentiating (2.1.17) with respect to t twice, multiplying the resulting equation by u_{tt} in $L^2(0,1)$, performing an integration by parts, using Theorems 2.1.1–2.1.2, we arrive at

$$\frac{1}{2}\frac{d}{dt}\|u_{tt}(t)\|^2 \leq -\lambda \int_0^1 \frac{|u_{tty}|^2}{v}\,dy + C_2 \Big(\|u_y\|_{L^4}^2 + \|\theta_{tt}\| + \|u_{ty}\| + \|\theta_t\|_{L^4}^2$$

$$+ \|\theta_t\|_{L^\infty}\|u_y\| + \|u_y\|_{L^\infty}\|u_{ty}\| + \|u_y\|_{L^6}^3 \Big)\|u_{tty}\|$$

$$\leq -C_1^{-1}\|u_{tty}\|^2 + C_2 \left(\|u_y\|_{H^2}^2 + \|\theta_t\|_{H^1}^2 + \|u_{ty}\|^2 + \|v_y\|^2 + \|\theta_{tt}\|^2 \right). \tag{2.4.27}$$

Thus, by Theorems 2.1.1–2.1.2, we deduce

$$\|u_{tt}(t)\|^2 + \int_0^t \|u_{tty}(\tau)\|^2\,d\tau$$

$$\leq \|u_{tt}(y,0)\|^2 + C \int_0^t \left(\|u_y\|_{H^2}^2 + \|\theta_t\|_{H^1}^2 + \|u_{ty}\|^2 + \|v_y\|^2 + \|\theta_{tt}\|^2 \right)(\tau)\,d\tau$$

$$\leq C_4 + C_2 \int_0^t \|\theta_{tt}(\tau)\|^2\,d\tau.$$

By (2.4.25), we conclude

$$\|u_{tt}(t)\|^2 + \int_0^t \|u_{tty}(\tau)\|^2\,d\tau \leq C_4 + C_2 \int_0^t \|\theta_{tyy}(\tau)\|^2\,d\tau$$

which, along with Theorems 2.1.1–2.1.2, gives estimate (2.4.3).

Differentiating (2.1.18) with respect to t twice, multiplying the resulting equation by $\vec{w}_{tt}$ in $L^2(0,1)$, performing an integration by parts, and using Theorems 2.1.1–2.1.2, we derive

$$\frac{1}{2}\frac{d}{dt}\|\vec{w}_{tt}(t)\|^2 \tag{2.4.28}$$

$$\leq -\mu \int_0^1 \frac{|\vec{w}_{tty}|^2}{v}\,dy + C_2 \Big(\|\vec{w}_{ty}\|_{L^\infty}\|v_y\| + \|\vec{w}_y\|_{L^\infty}\|u_{ty}\| + \|\vec{w}_y\|_{L^\infty}\|u_y\|_{L^4}^2 \Big)\|\vec{w}_{tty}\|$$

$$\leq -(2C)_1^{-1}\|\vec{w}_{tty}\|^2 + C_2 \left(\|\vec{w}_{tyy}\|^2 + \|v_y\|^2 + \|\vec{w}_y\|^2 + \|u_y\|_{H^1}^2 + \|u_{ty}\|^2 \right).$$

By (2.4.28), we can obtain

$$\|\vec{w}_{tt}(t)\|^2 + \int_0^t \|\vec{w}_{tty}(\tau)\|^2\,dy \leq C_4 + C_2 \int_0^t \|\vec{w}_{tyy}(\tau)\|^2\,d\tau$$

which, along with Theorems 2.1.1–2.1.2, gives estimate (2.4.4).

Differentiating (2.1.19) with respect to t twice, multiplying the resulting equation by θ_{tt} in $L^2(0,1)$, performing an integration by parts, and using Theorems 2.1.1–2.1.2, we infer that

$$\frac{1}{2}\frac{d}{dt}\int_0^1 e_\theta \theta_{tt}^2\, dy$$

$$= -\int_0^1 \left(\frac{\kappa\theta_y}{v}\right)_{tt}\theta_{tty}\, dy - \int_0^1 (e_{\theta tt}\theta_t + e_{utt}v_t)\,\theta_{tt}\, dy - \frac{3}{2}\int_0^1 e_\theta \theta_{tt}^2\, dy$$

$$- 2\int_0^1 \left[e_{vt} - \left(-p + \frac{\lambda u_y}{v}\right)_t\right]u_{ty}\theta_{tt}\, dy - \int_0^1 \left(e_v + p - \frac{\lambda u_y}{v}\right)u_{tty}\theta_{tt}\, dy$$

$$+ \int_0^1 \left(-p + \frac{\lambda u_y}{v}\right)_{tt} u_y\theta_{tt}\, dy + \mu\int_0^1 \left(\frac{\vec{w}_y\cdot\vec{w}}{v}\right)_{tt}\theta_{tt}\, dy$$

$$= A_1 + A_2 + A_3 + A_4 + A_5 + A_6 + A_7. \tag{2.4.29}$$

By virtue of Theorems 2.1.1–2.1.2, (2.4.1)–(2.4.2) and using the embedding theorem, we deduce that for any $\varepsilon \in (0,1)$,

$$A_1 = -\kappa\int_0^1 \left(\frac{\theta_y}{v}\right)_{tt}\theta_{tty}\, dy \tag{2.4.30}$$

$$= -\kappa\int_0^1 \frac{\theta_{tty}^2}{v}\, dy + \kappa\int_0^1 \left(\frac{2\theta_{ty}u_y}{v^2} + \frac{\theta_y u_{ty}}{v^2} - 2\frac{\theta_y u_y^2}{v^3}\right)\theta_{tty}\, dy$$

$$\le -C_1^{-1}\|\theta_{tty}\|^2 + C_2\|\theta_{tty}\|\left(\|\theta_{ty}\|_{L^\infty}\|u_y\| + \|\theta_y\|_{L^\infty}\|u_{ty}\| + \|\theta_y\|_{L^\infty}\|u_y\|_{L^4}^2\right)$$

$$\le -(2C_1)^{-1}\|\theta_{tty}\|^2 + C_2\left(\|\theta_{ty}\|^2 + \|\theta_y\|_{H^1}^2 + \|u_{ty}\|^2 + \|u_y\|_{H^1}^2 + \|\theta_{tyy}\|^2\right),$$

$$A_2 = -\int_0^1 (e_{\theta tt}\theta_t + e_{utt}v_t)\,\theta_{tt}\, dy \tag{2.4.31}$$

$$\le C_1\int_0^1 \left[(|v_t| + |\theta_t|)^2 + |v_{tt}| + |\theta_{tt}|\right](|\theta_t| + |v_t|)|\theta_{tt}|\, dy$$

$$\le C_1\|\theta_{tt}\|_{L^\infty}(\|\theta_t\| + \|u_y\|)[(\|u_y\|_{L^\infty} + \|\theta_t\|_{L^\infty})(\|\theta_t\| + \|u_y\|) + \|u_{ty}\| + \|\theta_{tt}\|]$$

$$\le \varepsilon\|\theta_{tty}\|^2 + C_2\varepsilon^{-1}(\|u_{ty}\|^2 + \|u_y\|_{H^1}^2 + \|\theta_{ty}\|^2 + \|\theta_{tt}\|^2 + \|\theta_t\|^2),$$

$$A_3 = -\frac{3}{2}\int_0^1 e_\theta\theta_{tt}^2\, dy \le C_1\int_0^1 (|\theta_t| + |u_y|)|\theta_{tt}|^2\, dy \le \varepsilon\|\theta_{tty}\|^2 + C_2\varepsilon^{-1}\|\theta_{tt}\|^2,$$

$$\tag{2.4.32}$$

$$A_4 = -2\int_0^1 \left[e_{vt} - \left(-p + \frac{\lambda u_y}{v}\right)_t\right]u_{ty}\theta_{tt}\, dy \tag{2.4.33}$$

$$\le C_1\int_0^1 (|\theta_t| + |u_y| + |u_{ty}| + |u_y^2|)\,|u_{ty}||\theta_{tt}|\, dy$$

$$\le C_2\|u_{ty}\|_{L^\infty}\left(\|u_y\| + \|\theta_t\| + \|u_{ty}\| + \|u_y\|_{L^4}^2\right)\|\theta_{tt}\|$$

$$\le C_2\|u_{ty}\|^{\frac{1}{2}}\|u_{tyy}\|^{\frac{1}{2}}\left(\|u_y\|_{H^1} + \|u_{ty}\| + \|\theta_t\|\right)\|\theta_{tt}\|$$

which implies

$$\int_0^t A_4(\tau)\,d\tau \le C_2 \sup_{0\le\tau\le t} \|\theta_{tt}(\tau)\| \int_0^t \|u_{ty}\|^{\frac{1}{2}} \|u_{tyy}\|^{\frac{1}{2}} \left(\|u_y\|_{H^1} + \|u_{ty}\| + \|\theta_t\|\right) d\tau$$

$$\le C_2 \sup_{0\le\tau\le t} \|\theta_{tt}(\tau)\| \left(\int_0^t \|u_{ty}(\tau)\|^2\,d\tau\right)^{\frac{1}{4}} \left(\int_0^t \|u_{tyy}(\tau)\|^2\,d\tau\right)^{\frac{1}{4}}$$

$$\times \left(\int_0^t (\|u_y\|_{H^1}^2 + \|u_{ty}\|^2 + \|\theta_t\|^2)(\tau)\,d\tau\right)^{\frac{1}{2}}$$

$$\le \varepsilon \left(\sup_{0\le\tau\le t} \|\theta_{tt}(\tau)\|^2 + \int_0^t \|u_{tyy}(\tau)\|^2\,d\tau\right) + C_2\varepsilon^{-3}, \qquad (2.4.34)$$

$$A_5 = -\int_0^1 \left(e_v + p - \frac{\lambda u_y}{v}\right) u_{tty}\theta_{tt}\,dy \le \varepsilon\|u_{tty}\|^2 + C_2\varepsilon^{-1}\|\theta_{tt}\|^2, \tag{2.4.35}$$

$$A_6 = \int_0^1 \left(-p + \frac{\lambda u_y}{v}\right)_{tt} u_y\theta_{tt}\,dy$$

$$\le C_1 \int_0^1 \left((|v_t| + |\theta_t|)^2 + |\theta_{tt}| + |u_{tty}| + |u_{ty}u_y| + |u_y|^3\right)|u_y||\theta_{tt}|\,dy$$

$$\le C_1\|u_y\|_{L^\infty}\|\theta_{tt}\| \Big((\|v_t\|_{L^\infty} + \|\theta_t\|_{L^\infty})(\|v_t\| + \|\theta_t\|) + \|u_{tty}\| + \|u_y\|$$

$$+ \|\theta_{tt}\| + \|u_{ty}\| + \|u_y\|_{L^\infty} + \|u_y\|_{L^6}^3\Big)$$

$$\le \varepsilon\|u_{tty}\|^2 + C_2\varepsilon^{-1}\left(\|u_{tt}\|^2 + \|u_y\|^2 + \|\theta_{tt}\|^2 + \|\theta_{ty}\|^2 + \|\theta_t\|^2\right), \tag{2.4.36}$$

$$A_7 = \mu \int_0^1 \left(\frac{\vec{w}_y \cdot \vec{w}}{v}\right)_{tt} \theta_{tt}\,dy$$

$$\le C_2\|\theta_{tt}\| \Big(\|\vec{w}_{ty}\|_{L^4}^2 + \|\vec{w}_{tty}\|_{L^\infty} + \|\vec{w}_y\|_{L^\infty}\|\vec{w}_{ty}\|\|u_y\|_{L^\infty}$$

$$+ \|\vec{w}_y\|_{L^\infty}^2\|u_{yy}\| + \|u_y\|_{L^\infty}^2\|\vec{w}_y\|_{L^4}^2\Big)$$

$$\le C_2\varepsilon^{-1}\|\theta_{tt}\|^2 + \varepsilon\Big(\|\vec{w}_{tyy}\|^2 + \|\vec{w}_{ty}\|^2$$

$$+ \|\vec{w}_{tty}\|^2 + \|\vec{w}_y\|_{H^1}^2 + \|u_y\|_{H^1}^2\Big). \tag{2.4.37}$$

Integrating with respect to t, using Theorems 2.1.1–2.1.2 and (2.4.35)–(2.4.37), we can derive

$$\|\sqrt{e_\theta}\theta_{tt}\|^2 \le C_4 + \int_0^t (A_1 + A_2 + A_3 + A_4 + A_5 + A_6 + A_7)(\tau)\,d\tau.$$

Thus

$$\|\theta_{tt}(t)\|^2 + \int_0^t \|\theta_{tty}(\tau)\|^2 \, d\tau$$

$$\leq C_1 \varepsilon \left\{ \sup_{0 \leq \tau \leq t} \|\theta_{tt}(\tau)\|^2 + \int_0^t (\|u_{tty}\|^2 + \|u_{tyy}\|^2)(\tau) \, d\tau \right\}$$

$$+ C_4 \varepsilon^{-3} + C_2 \varepsilon^{-1} \int_0^t (\|\theta_{tt}\|^2 + \|\theta_{tty}\|^2)(\tau) \, d\tau$$

$$+ C_1 \varepsilon \int_0^t (\|\vec{w}_{tyy}\|^2 + \|\vec{w}_{tty}\|^2)(\tau) \, d\tau \qquad (2.4.38)$$

which, along with Theorems 2.1.1–2.1.2 and (2.4.4), (2.4.25), (2.4.38), gives estimate (2.4.5). □

Lemma 2.4.2. *Under assumptions of Theorem 2.1.3, for any $(v_0, u_0, \vec{w}_0, \theta_0) \in H_+^4$ and for any $\varepsilon \in (0,1)$, we have for any $t > 0$,*

$$\|u_{ty}(t)\|^2 + \int_0^t \|u_{tyy}(\tau)\|^2 \, d\tau \leq C_3 \varepsilon^{-6} + C_1 \varepsilon^2 \int_0^t (\|u_{tty}\|^2 + \|\theta_{tyy}\|^2)(\tau) \, d\tau,$$
$$\qquad (2.4.39)$$

$$\|\vec{w}_{ty}(t)\|^2 + \int_0^t \|\vec{w}_{tyy}(\tau)\|^2 \, d\tau \leq C_3 \varepsilon^{-6} + C_1 \varepsilon^2 \int_0^t \|\vec{w}_{tty}(\tau)\|^2 \, d\tau, \qquad (2.4.40)$$

$$\|\theta_{ty}(t)\|^2 + \int_0^t \|\theta_{tyy}(\tau)\|^2 \, d\tau$$

$$\leq C_3 \varepsilon^{-6} + C_2 \varepsilon^2 \int_0^t (\|u_{tyy}\|^2 + \|\vec{w}_{tyy}\|^2 + \|\theta_{tty}\|^2)(\tau) \, d\tau. \qquad (2.4.41)$$

Proof. Differentiating (2.1.17) with respect to y and t, multiplying the resulting equation by u_{ty} and integrating by parts in $L^2(0,1)$, we arrive at

$$\frac{1}{2} \frac{d}{dt} \|u_{ty}(t)\|^2 = \left(-p + \frac{\lambda u_y}{v} \right)_{ty} u_{ty} \Big|_{y=0}^{y=1} - \int_0^1 \left(-p + \frac{\lambda u_y}{v} \right)_{ty} u_{tyy} \, dy \qquad (2.4.42)$$

$$= B_1 + B_2.$$

We employ Theorems 2.1.1–2.1.2 and Lemma 2.4.1, the interpolation inequality and Poincaré's inequality to get

$$B_1 \leq C_1 \Big(\|u_{yy}\|_{L^\infty} + (\|u_y\|_{L^\infty} + \|\theta_t\|_{L^\infty})(\|v_y\|_{L^\infty} + \|\theta_y\|_{L^\infty})$$

$$+ \|\theta_{ty}\|_{L^\infty} + \|u_{tyy}\|_{L^\infty} + \|u_{ty}\|_{L^\infty} \|v_y\|_{L^\infty} + \|u_{yy}\|_{L^\infty} \|u_y\|_{L^\infty}$$

$$+ \|u_y\|_{L^\infty}^2 \|v_y\|_{L^\infty} \Big) \|u_{ty}\|_{L^\infty}$$

$$\leq C_2 (B_{11} + B_{12}) \|u_{ty}\|^{\frac{1}{2}} \|u_{tyy}\|^{\frac{1}{2}} \qquad (2.4.43)$$

where

$$B_{11} = \|v_y\|_{H^2} + \|\theta_t\| + \|\theta_{ty}\|,$$

$$B_{12} = \|u_{ty}\|^{\frac{1}{2}} \|u_{tyy}\|^{\frac{1}{2}} + \|u_{tyyy}\|^{\frac{1}{2}} \|u_{tyy}\|^{\frac{1}{2}} + \|\theta_{ty}\|^{\frac{1}{2}} \|\theta_{tyy}\|^{\frac{1}{2}}.$$

Applying Young's inequality several times, we have that for any $\varepsilon \in (0,1)$,

$$C_2 B_{11} \|u_{ty}\|^{\frac{1}{2}} \|u_{tyy}\|^{\frac{1}{2}} \leq \frac{\varepsilon^2}{2} \|u_{tyy}\|^2 + C_2 \varepsilon^{-\frac{2}{3}} \left(\|u_{ty}\|^2 + \|u_y\|_{H^2}^2 + \|\theta_t\|^2 + \|\theta_{ty}\|^2 \right),$$
$$(2.4.44)$$

$$C_2 B_{11} \|u_{ty}\|^{\frac{1}{2}} \|u_{tyy}\|^{\frac{1}{2}} \leq \frac{\varepsilon^2}{2} \|u_{tyy}\|^2 + \varepsilon^2 \left(\|\theta_{tyy}\|^2 + \|u_{tyyy}\|^2 \right)$$
$$+ C_2 \varepsilon^{-6} \left(\|\theta_{ty}\|^2 + \|u_{ty}\|^2 \right). \qquad (2.4.45)$$

Thus we infer from (2.4.43)–(2.4.45) and Theorems 2.1.1–2.1.2 and Lemma 2.4.1,

$$B_1 \leq \varepsilon^2 (\|u_{tyy}\|^2 + \|\theta_{tyy}\|^2 + \|u_{tyyy}\|^2) + C_2 \varepsilon^{-6} (\|u_{ty}\|^2 + \|u_y\|_{H^2}^2 + \|\theta_t\|^2 + \|\theta_{ty}\|^2)$$
$$(2.4.46)$$

which, together with Lemmas 2.2.1–2.2.3, further leads to

$$\int_0^t B_1(\tau) \, d\tau \leq \varepsilon^2 \int_0^t \left(\|u_{tyy}\|^2 + \|\theta_{tyy}\|^2 + \|u_{tyyy}\|^2 \right)(\tau) \, d\tau + C_2 \varepsilon^{-6}, \ \forall t > 0,$$
$$(2.4.47)$$

$$B_2 \leq -\lambda \int_0^1 \frac{u_{tyy}^2}{v} \, dy + C_1 \Big[(\|u_y\| + \|\theta_t\|)(\|v_y\|_{L^\infty} + \|\theta_y\|_{L^\infty}) + \|u_{yy}\| + \|\theta_{ty}\|$$

$$+ \|v_y\|_{L^\infty} \|u_{ty}\| + \|u_y\|_{L^\infty} \|u_{yy}\| + \|u_y\|_{L^\infty}^2 \|v_y\| \Big] \|u_{tyy}\|$$

$$\leq (-2C_1)^{-1} \|u_{tyy}\|^2 + C_2 \varepsilon^2 \left(\|u_{ty}\|^2 + \|u_y\|_{H^1}^2 + \|v_y\|_{H^1}^2 + \|\theta_t\|_{H^1}^2 \right). \quad (2.4.48)$$

By (2.4.42), (2.4.47)–(2.4.48) and Theorems 2.1.1–2.1.2 and Lemma 2.4.1, for any $\varepsilon \in (0,1)$ small enough, we have

$$\|u_{ty}(t)\|^2 + \int_0^t \|u_{tyy}(\tau)\|^2 \, d\tau \leq C_2 \varepsilon^{-6} + C_1 \varepsilon^2 \int_0^t (\|u_{tyyy}\|^2 + \|\theta_{tyy}\|^2)(\tau) \, d\tau.$$
$$(2.4.49)$$

On the other hand, differentiating (2.1.17) with respect to y and t, and using Theorems 2.1.1–2.1.2 and Lemma 2.4.1, we derive

$$\|u_{tyyy}(t)\| \leq C_2 \left(\|u_{tty}\| + \|u_{tyy}\| + \|\theta_t\|_{H^2} + \|\theta_y\|_{H^1} + \|u_y\|_{H^2} + \|u_y\|_{H^1} \right).$$
$$(2.4.50)$$

Thus inserting (2.4.50) into (2.4.49) implies estimate (2.4.39).

Differentiating (2.1.18) with respect to y and t, multiplying the resulting equation by $\vec{w}_{ty}$ in $L^2(0,1)$, and integrating by parts, we can obtain

$$\frac{1}{2} \frac{d}{dt} \|\vec{w}_{ty}(t)\|^2 = \mu \left(\frac{\vec{w}_y}{v} \right)_{ty} \vec{w}_{ty} \Big|_{y=0}^{y=1} - \mu \int_0^1 \left(\frac{\vec{w}_y}{v} \right)_{ty} \vec{w}_{tyy} \, dy$$
$$= D_1 + D_2$$
$$(2.4.51)$$

where

$$D_1 \leq C_1 \|\vec{w}_{ty}\|_{L^\infty} \Big(\|\vec{w}_{tyy}\|_{L^\infty} + \|\vec{w}_{ty}\|_{L^\infty} \|v_y\|_{L^\infty} + \|\vec{w}_{yy}\|_{L^\infty} \|u_y\|_{L^\infty}$$
$$+ \|\vec{w}_y\|_{L^\infty} \|u_{yy}\|_{L^\infty} + \|\vec{w}_y\|_{L^\infty} \|v_y u_y\|_{L^\infty} \Big)$$
$$\leq C_1 \varepsilon^2 (\|\vec{w}_{tyy}\|^2 + \|\vec{w}_{tyyy}\|^2) + C_2 \varepsilon^{-6} \Big(\|\vec{w}_{ty}\|^2 + \|\vec{w}_y\|_{H^2}^2$$
$$+ \|u_y\|_{H^2}^2 + \|v_y\|_{H^1}^2 \Big), \tag{2.4.52}$$

$$D_2 \leq -\mu \int_0^1 \frac{|\vec{w}_{tyy}|^2}{v} \, dy + C_1 \|\vec{w}_{tyy}\| \Big(\|\vec{w}_{ty}\|_{L^\infty} \|v_y\| + \|\vec{w}_{yy}\| \|u_y\|_{L^\infty}$$
$$+ \|\vec{w}_y\|_{L^\infty} \|u_{yy}\| + \|\vec{w}_{ty}\|_{L^\infty} \|v_y\| \|u_y\|_{L^\infty} \Big)$$
$$\leq -(2C_1)^{-1} \|\vec{w}_{tyy}\|^2 + C_2 \varepsilon^2 \Big(\|\vec{w}_{ty}\|^2 + \|v_y\|^2 + \|u_y\|_{H^1}^2$$
$$+ \|\vec{w}_y\|_{H^1}^2 + \|\vec{w}_{tyy}\|^2 \Big). \tag{2.4.53}$$

By (2.4.51)–(2.4.53) and Theorems 2.1.1–2.1.2 and Lemma 2.4.1, we have

$$\|\vec{w}_{ty}(t)\|^2 + \int_0^t \|\vec{w}_{tyy}(\tau)\|^2 \, d\tau \leq C_2 \varepsilon^{-6} + C_1 \varepsilon^2 \int_0^t \|\vec{w}_{tyyy}(\tau)\|^2 \, d\tau. \tag{2.4.54}$$

Differentiating (2.1.18) with respect to y and t, and using Lemmas 2.2.1–2.2.4, we derive

$$\|\vec{w}_{tyyy}(t)\| \leq C_2 (\|\vec{w}_{tty}\| + \|\vec{w}_{tyy}\| + \|\vec{w}_y\|_{H^2} + \|u_y\|_{H^2} + \|v_y\|_{H^2} + \|\vec{w}_{ty}\|). \tag{2.4.55}$$

Thus inserting (2.4.55) into (2.4.54) implies estimate (2.4.40).

Analogously, we get from (2.1.19),

$$\frac{1}{2} \frac{d}{dt} \int_0^1 e_\theta \theta_{ty}^2 \, dy = E_1 + E_2 + E_3 + E_4 + E_5 \tag{2.4.56}$$

where

$$E_1 = \kappa \left(\frac{\theta_y}{v} \right)_{ty} \theta_{ty} \Big|_{y=0}^{y=1},$$

$$E_2 = -\kappa \int_0^1 \left(\frac{\theta_y}{v} \right)_{ty} \theta_{tyy} \, dy,$$

$$E_3 = -\int_0^1 \left[\left(e_v - p + \frac{\lambda u_y}{v} \right) u_y \right]_{ty} \theta_{ty} \, dy,$$

$$E_4 = -\int_0^1 \left(e_{\theta ty} \theta_t + e_{\theta y} \theta_{tt} + \frac{1}{2} e_{\theta t} \theta_{ty} \right) \theta_{ty} \, dy,$$

$$E_5 = \mu \int_0^1 \left(\frac{\vec{w}_y \cdot \vec{w}}{v} \right)_{ty} \theta_{ty} \, dy.$$

It follows that

$$E_1 \leq C_2 \|\theta_{ty}\|_{L^\infty} \Big(\|\theta_{tyy}\|_{L^\infty} + \|\theta_{ty}\|_{L^\infty}\|v_y\|_{L^\infty} + \|\theta_{yy}\|_{L^\infty}\|u_y\|_{L^\infty}$$

$$+ \|\theta_y\|_{L^\infty}\|u_{yy}\|_{L^\infty} + \|\theta_y\|_{L^\infty}\|u_y\|_{L^\infty}\|v_y\|_{L^\infty} \Big)$$

$$\leq C_2 \|\theta_{ty}\|^{\frac12}\|\theta_{tyy}\|^{\frac12} \Big(\|\theta_{tyy}\|^{\frac12}\|\theta_{tyyy}\|^{\frac12} + \|\theta_{ty}\|^{\frac12}\|\theta_{tyy}\|^{\frac12}$$

$$+ \|\theta_y\|_{H^2} + \|u_y\|_{H^2} + \|v_y\|_{H^1} \Big),$$

$$\|\theta_{ty}\|^{\frac12}\|\theta_{tyy}\|^{\frac12}\|\theta_{tyy}\|^{\frac12}\|\theta_{tyyy}\|^{\frac12} \leq \frac{\varepsilon^2}{3}\big(\|\theta_{tyy}\|^2 + \|\theta_{tyyy}\|^2 \big) + C_2\varepsilon^{-6}\|\theta_{ty}\|^2,$$

$$\Big(\|\theta_{ty}\|^{\frac12}\|\theta_{tyy}\|^{\frac12} \Big)^2 \leq \frac{\varepsilon^2}{3}\|\theta_{tyy}\|^2 + C_2\varepsilon^{-2}\|\theta_{ty}\|^2,$$

$$\|\theta_{ty}\|^{\frac12}\|\theta_{tyy}\|^{\frac12}\big(\|\theta_y\|_{H^2} + \|u_y\|_{H^2} + \|v_y\|_{H^1} \big)$$

$$\leq \frac{\varepsilon^2}{3}\|\theta_{tyy}\|^2 + C_2\varepsilon^{-2}\Big(\|\theta_{ty}\|^2 + \|\theta_y\|_{H^2}^2 + \|u_y\|_{H^2}^2 + \|v_y\|_{H^1}^2 \Big).$$

Thus

$$E_1 \leq \varepsilon^2(\|\theta_{tyy}\|^2 + \|\theta_{tyyy}\|^2) + C_2\varepsilon^{-6}\Big(\|\theta_{ty}\|^2 + \|\theta_y\|_{H^2}^2$$

$$+ \|u_y\|_{H^2}^2 + \|v_y\|_{H^1}^2 \Big), \tag{2.4.57}$$

$$E_2 \leq -\kappa \int_0^1 \frac{\theta_{tyy}^2}{v}\, dy + C_2\|\theta_{tyy}\| \Big(\|\theta_{ty}\|\|v_y\|_{L^\infty} + \|\theta_{yy}\|\|u_y\|_{L^\infty}$$

$$+ \|u_{yy}\|\|\theta_y\|_{L^\infty} + \|u_y v_y\|\|\theta_y\|_{L^\infty} \Big)$$

$$\leq -(2C_1)^{-1}\|\theta_{tyy}\|^2 + C_2\varepsilon^2\Big(\|\theta_{ty}\|^2 + \|\theta_y\|_{H^1}^2 + \|u_y\|_{H^1}^2 \Big). \tag{2.4.58}$$

Similarly,

$$E_3 \leq \varepsilon^2\|u_{tyy}\|^2 + C_2\varepsilon^{-2}\big(\|u_y\|_{H^2}^2 + \|u_{ty}\|^2 + \|\theta_t\|_{H^1}^2 \big), \tag{2.4.59}$$

$$E_4 \leq \varepsilon^2\|\theta_{tyy}\|^2 + C_2\varepsilon^{-2}\Big(\|u_y\|_{H^1}^2 + \|u_{ty}\|^2 + \|\theta_t\|_{H^1}^2$$

$$+ \|\theta_y\|_{H^2}^2 + \|u_{ty}\|^2 \Big), \tag{2.4.60}$$

$$E_5 \leq \varepsilon^2\|\vec{w}_{tyy}\|^2 + C_2\varepsilon^2\|\theta_{tyy}\|^2 + C_2\varepsilon^{-2}\big(\|\vec{w}_y\|_{H^2}^2 + \|\vec{w}_{ty}\|^2 + \|u_y\|_{H^1}^2 \big). \tag{2.4.61}$$

By (2.4.56)–(2.4.61) and Theorems 2.1.1–2.1.2 and Lemma 2.4.1, we have

$$\|\theta_{ty}(t)\|^2 + \int_0^t \|\theta_{tyy}(\tau)\|^2\, d\tau \leq C_1\varepsilon^2 \int_0^t (\|\theta_{tyyy}\|^2 + \|u_{tyy}\|^2 + \|\vec{w}_{tyy}\|^2)(\tau)\, d\tau$$

$$+ C_2\varepsilon^{-6}\int_0^t (\|\theta_{tyy}\|^2 + \|\theta_y\|_{H^2}^2 + \|u_y\|_{H^2}^2 + \|v_y\|_{H^1}^2)(\tau)\, d\tau$$

$$+ C_2\varepsilon^{-2}\int_0^t (\|\vec{w}_y\|_{H^2}^2 + \|\vec{w}_{ty}\|^2)(\tau)\, d\tau. \tag{2.4.62}$$

Differentiating (2.1.19) with respect to y and t, and using Theorems 2.1.1–2.1.2 and Lemma 2.4.1, we derive

$$\|\theta_{tyyy}\| \le C_2 \Big(\|u_{ty}\|_{H^1} + \|\theta_t\|_{H^1} + \|\theta_y\|_{H^2} + \|\theta_{tt}\|_{H^1} + \|\vec{w}_{ty}\|_{H^1}$$

$$+ \|\vec{w}_y\|_{H^1} + \|v_y\|_{H^1} \Big). \tag{2.4.63}$$

By (2.4.62)–(2.4.63) and Theorems 2.1.1–2.1.2 and Lemma 2.4.1, we derive estimate (2.4.41). $\qquad\square$

Lemma 2.4.3. *Under assumptions of Theorem 2.1.3, for any $(v_0, u_0, \vec{w}_0, \theta_0) \in H_+^4$ and for any $\varepsilon \in (0,1)$, we have for any $t > 0$,*

$$\|u_{tt}(t)\|^2 + \|\vec{w}_{tt}(t)\|^2 + \|\theta_{tt}(t)\|^2 + \|u_{ty}(t)\|^2 + \|\vec{w}_{ty}(t)\|^2 + \|\theta_{ty}(t)\|^2 \tag{2.4.64}$$

$$+ \int_0^t \Big(\|u_{tty}\|^2 + \|u_{tyy}\|^2 + \|\vec{w}_{tty}\|^2 + \|\vec{w}_{tyy}\|^2 + \|\theta_{tty}\|^2 + \|\theta_{tyy}\|^2 \Big)(\tau)\, d\tau \le C_4,$$

$$\|u_{yyy}(t)\|_{H^1}^2 + \|u_{yy}(t)\|_{W^{1,\infty}}^2 + \|\theta_{yyy}(t)\|_{H^1}^2 + \|\theta_{yy}(t)\|_{w^{1,\infty}}^2 + \|\vec{w}_{yyy}(t)\|_{H^1}^2$$

$$+ \|\vec{w}_{yy}(t)\|_{W^{1,\infty}}^2 + \|v_{tyyy}(t)\|^2 + \|u_{tyy}(t)\|^2 + \|\vec{w}_{tyy}(t)\|^2 + \|\theta_{tyy}(t)\|^2$$

$$+ \int_0^t \Big(\|u_{tt}\|^2 + \|\vec{w}_{tt}\|^2 + \|\theta_{yy}\|_{w^{1,\infty}}^2 + \|\theta_{tt}\|^2 + \|u_{yy}\|_{W^{1,\infty}}^2 + \|\vec{w}_{yy}\|_{W^{1,\infty}}^2$$

$$+ \|u_{tyy}\|_{H^1}^2 + \|\theta_{tyy}\|_{H^1}^2 + \|\vec{w}_{tyy}\|_{H^1}^2 + \|u_{ty}\|_{W^{1,\infty}}^2 + \|\theta_{ty}\|_{w^{1,\infty}}^2$$

$$+ \|\vec{w}_{ty}\|_{W^{1,\infty}}^2 + \|v_{tyyy}\|_{H^1}^2 \Big)(\tau)\, d\tau \le C_4, \tag{2.4.65}$$

$$\|v_{yyy}(t)\|_{H^1}^2 + \|v_{yy}(t)\|_{W^{1,\infty}}^2 + \int_0^t \Big(\|v_{yyy}\|_{H^1}^2 + \|v_{yy}\|_{W^{1,\infty}}^2 \Big)(\tau)\, d\tau \le C_4, \tag{2.4.66}$$

$$\int_0^t \Big(\|u_{yyyy}\|_{H^1}^2 + \|\theta_{yyyy}\|_{H^1}^2 + \|\vec{w}_{yyyy}\|_{H^1}^2 \Big)(\tau)\, d\tau \le C_4. \tag{2.4.67}$$

Proof. Adding up (2.4.39), (2.4.40) and (2.4.41), picking $\varepsilon \in (0,1)$ small enough, we arrive at

$$\|u_{ty}(t)\|^2 + \|\vec{w}_{ty}(t)\|^2 + \|\theta_{ty}(t)\|^2 + \int_0^t \Big(\|u_{tyy}\|^2 + \|\vec{w}_{tyy}\|^2 + \|\theta_{tyy}\|^2 \Big)(\tau)\, d\tau$$

$$\le C_3 \varepsilon^{-6} + C_1 \varepsilon^2 \int_0^t \Big(\|u_{tty}\|^2 + \|\vec{w}_{tty}\|^2 + \|\theta_{tty}\|^2 \Big)(\tau)\, d\tau. \tag{2.4.68}$$

Now multiplying (2.4.3) and (2.4.4) by ε respectively, multiplying (2.4.5) by $\varepsilon^{3/2}$, then adding the resultant to (2.4.68), and choosing $\varepsilon \in (0,1)$ small enough, we obtain (2.4.64).

Differentiating (2.1.17) with respect to y, and using (2.1.16), we derive

$$\lambda \frac{\partial}{\partial t} \left(\frac{v_{yy}}{v} \right) - p_v v_{yy} = u_{ty} + E(y,t) \tag{2.4.69}$$

where

$$E(y,t) = 2\lambda \frac{u_{yy}u_y}{v^2} - 2\lambda \frac{u_y v_y^2}{v^3} + p_{vv}v_y^2 + 2p_{v\theta}v_y\theta_y + p_{\theta\theta}\theta_y^2 + p_\theta\theta_{yy}.$$

Differentiating (2.4.69) with respect to y, we have

$$\lambda \frac{\partial}{\partial t}\left(\frac{v_{yyy}}{v}\right) - p_v v_{yyy} = E_1(y,t) \tag{2.4.70}$$

where

$$E_1(y,t) = E_y(y,t) + p_{vy}v_{yy} + u_{tyy} + \lambda\left(\frac{v_{yy}v_y}{v^2}\right)_t,$$
$$\|E_1(t)\| \leq C_2\left(\|u_{tyy}\| + \|u_y\|_{H^2} + \|v_y\|_{H^1} + \|\theta_y\|_{H^2}\right).$$

Thus

$$\int_0^t \|E_1(\tau)\|^2 d\tau \leq C_4. \tag{2.4.71}$$

Multiplying (2.4.70) by $\frac{v_{yyy}}{v}$ in $L^2(0,1)$, we can obtain

$$\frac{d}{dt}\left\|\frac{v_{yyy}}{v}\right\|^2 + C_1^{-1}\left\|\frac{v_{yyy}}{v}\right\|^2 \leq C_1\|E_1(t)\|^2. \tag{2.4.72}$$

We infer from (2.4.71)–(2.4.72) that

$$\|v_{yyy}(t)\|^2 + \int_0^t \|v_{yyy}(\tau)\|^2 d\tau \leq C_4, \quad \forall t > 0 \tag{2.4.73}$$

which, together with (2.4.9), (2.4.15), (2.4.20) and (2.4.64), gives

$$\|u_{yyy}(t)\| + \|\vec{w}_{yyy}(t)\| + \|\theta_{yyy}(t)\| \leq C_4. \tag{2.4.74}$$

By (2.4.12), (2.4.17), (2.4.22) and (2.4.64), we have

$$\int_0^t (\|u_{yyy}\|_{H^1}^2 + \|\theta_{yyy}\|_{H^1}^2 + \|\vec{w}_{yyy}\|_{H^1}^2)(\tau)\,d\tau \leq C_4. \tag{2.4.75}$$

Using the embedding theorem and (2.4.73)–(2.4.75), we have

$$\|u_{yyy}(t)\|^2 + \|\vec{w}_{yyy}(t)\|^2 + \|\theta_{yyy}(t)\|^2 + \|u_{yy}(t)\|_{L^\infty}^2 + \|\theta_{yy}(t)\|_{L^\infty}^2 + \|\vec{w}_{yy}(t)\|_{L^\infty}^2$$
$$+ \int_0^t (\|u_{yyy}\|_{H^1}^2 + \|\theta_{yyy}\|_{H^1}^2 + \|\vec{w}_{yyy}\|_{H^1}^2)(\tau)\,d\tau \leq C_4. \tag{2.4.76}$$

Differentiating (2.1.17), (2.1.19) with respect to t respectively, and using Lemmas 2.4.1–2.4.2, we have

$$\|u_{tyy}(t)\| \leq C_1\|u_{tt}(t)\| + C_2(\|u_y(t)\|_{H^1} + \|v_y(t)\|_{H^1} + \|u_{ty}(t)\| + \|\theta_t(t)\|)$$
$$\leq C_4, \tag{2.4.77}$$

$$\|\vec{w}_{tyy}(t)\| \leq C_1\|\vec{w}_{tt}(t)\| + C_2(\|\vec{w}_y(t)\|_{H^1} + \|u_y(t)\|_{H^1} + \|v_y(t)\| + \|\vec{w}_{ty}(t)\|)$$
$$\leq C_4, \tag{2.4.78}$$

$$\|\theta_{tyy}(t)\| \leq C_1\|\theta_{tt}(t)\| + C_2(\|\theta_y(t)\|_{H^1} + \|u_y(t)\|_{H^1} + \|u_{ty}(t)\| + \|\theta_t(t)\|_{H^1}$$
$$+ \|\vec{w}_{ty}(t)\| + \|\vec{w}_y(t)\|_{H^1}) \leq C_4. \tag{2.4.79}$$

By (2.4.12), (2.4.17), (2.4.22), (2.4.75)–(2.4.79) and Lemmas 2.4.1–2.4.2, for any $t > 0$, we get

$$\|u_{yyyy}(t)\|^2 + \|\vec{w}_{yyyy}(t)\|^2 + \|\theta_{yyyy}(t)\|^2 \tag{2.4.80}$$
$$+ \int_0^t \left(\|u_{tyy}\|^2 + \|\vec{w}_{tyy}\|^2 + \|\theta_{tyy}\|^2 + \|u_{yyyy}\|^2 + \|\vec{w}_{yyyy}\|^2 + \|\theta_{yyyy}\|^2\right)(\tau)\,d\tau \leq C_4$$

which, together with (2.4.76) and the interpolation inequality, gives

$$\|u_{yyy}(t)\|_{L^\infty}^2 + \|\vec{w}_{yyy}(t)\|_{L^\infty}^2 + \|\theta_{yyy}(t)\|_{L^\infty}^2 \tag{2.4.81}$$
$$+ \int_0^t \left(\|u_{yyy}\|_{L^\infty}^2 + \|\vec{w}_{yyy}\|_{L^\infty}^2 + \|\theta_{yyy}\|_{L^\infty}^2\right)(\tau)\,d\tau \leq C_4.$$

Differentiating (2.4.70) with respect to y, we get

$$\lambda\frac{\partial}{\partial t}\left(\frac{v_{yyyy}}{v}\right) - p_v v_{yyyy} = E_2(t) \tag{2.4.82}$$

where

$$E_2(t) = E_{1y}(t) + p_{vy}v_{yyy} + \lambda\left(\frac{v_{yyy}v_y}{v^2}\right)_t.$$

By Lemmas 2.4.1–2.4.2, (2.4.77)–(2.4.80) and the embedding theorem, the interpolation inequality, we deduce

$$\|E_{1y}(t)\| \leq C_1\|u_{tyyy}\| + C_4\left(\|u_y\|_{H^3} + \|v_y\|_{H^2} + \|\theta_y\|_{H^3}\right),$$
$$\|E_2(t)\| \leq C_1\|u_{tyyy}\| + C_4\left(\|u_y\|_{H^3} + \|v_y\|_{H^2} + \|\theta_y\|_{H^3}\right). \tag{2.4.83}$$

By (2.4.23), (2.4.24), (2.4.26) and (2.4.80), we have

$$\int_0^t \left(\|u_{tt}\|^2 + \|\vec{w}_{tt}\|^2 + \|\theta_{tt}\|^2\right)(\tau)\,d\tau \leq C_4, \quad \forall t > 0 \tag{2.4.84}$$

which, together with (2.4.50), (2.4.55), (2.4.63) and (2.4.64), gives

$$\int_0^t \left(\|u_{tyyy}\|^2 + \|\vec{w}_{tyyy}\|^2 + \|\theta_{tyyy}\|^2\right)(\tau)\,d\tau \leq C_4, \quad \forall t > 0. \tag{2.4.85}$$

Thus

$$\int_0^t \|E_2(\tau)\|^2 \, d\tau \leq C_4, \quad \forall t > 0. \tag{2.4.86}$$

Multiplying (2.4.82) by $\frac{v_{yyyy}}{v}$ in $L^2(0,1)$, we can obtain

$$\frac{d}{dt} \left\| \frac{v_{yyyy}}{v} \right\|^2 + C_1^{-1} \left\| \frac{v_{yyyy}}{v} \right\|^2 \leq C_1 \|E_2(t)\|^2. \tag{2.4.87}$$

Thus

$$\left\| \frac{v_{yyyy}}{v}(t) \right\|^2 + \int_0^t \left\| \frac{v_{yyyy}}{v}(\tau) \right\|^2 \, d\tau \leq C_4, \quad \forall t > 0. \tag{2.4.88}$$

Differentiating (2.1.17), (2.1.18), (2.1.19) with respect to y three times respectively, using Lemmas 2.4.1–2.4.2 and Poincaré's inequality, we infer

$$\|u_{yyyyy}(t)\| \leq C_1 \|u_{tyyy}(t)\| + C_2 \left(\|u_y(t)\|_{H^3} + \|v_y(t)\|_{H^3} + \|\theta_y(t)\|_{H^3} \right), \tag{2.4.89}$$

$$\|\vec{w}_{yyyyy}(t)\| \leq C_1 \|\vec{w}_{tyyy}(t)\| + C_2 \left(\|\vec{w}_y(t)\|_{H^3} + \|v_y(t)\|_{H^3} \right), \tag{2.4.90}$$

$$\|\theta_{yyyyy}(t)\| \leq C_1 \|\theta_{tyyy}(t)\| + C_2 \big(\|u_y(t)\|_{H^3} + \|\vec{w}_y(t)\|_{H^3} + \|\theta_y(t)\|_{H^3}$$
$$+ \|\theta_{tyy}(t)\| + \|v_y(t)\|_{H^3} \big). \tag{2.4.91}$$

Using (2.1.16), (2.4.76), (2.4.85), (2.4.88)–(2.4.91) and Lemmas 2.4.1–2.4.2, the interpolation inequality, we have

$$\int_0^t \left(\|u_{yyyyy}\|^2 + \|v_{tyyyy}\|^2 + \|\vec{w}_{yyyyy}\|^2 + \|\theta_{yyyyy}\|^2 \right)(\tau) \, d\tau \leq C_4, \quad \forall t > 0, \tag{2.4.92}$$

$$\int_0^t \left(\|u_{yy}\|_{W^{2,\infty}}^2 + \|\theta_{yy}\|_{W^{2,\infty}}^2 + \|\vec{w}_{yy}\|_{W^{2,\infty}}^2 \right)(\tau) \, d\tau \leq C_4, \quad \forall t > 0. \tag{2.4.93}$$

Finally, using (2.1.16), (2.4.76)–(2.4.81), (2.4.84)–(2.4.85), (2.4.92)–(2.4.93) and Sobolev's interpolation inequality, we can derive the desired estimates (2.4.65)–(2.4.67). □

Lemma 2.4.4. *Under assumptions of Theorem 2.1.3, for any* $(v_0, u_0, \vec{w}_0, \theta_0) \in H_+^4$ *and for any* $\varepsilon \in (0,1)$, *we have for any* $t > 0$,

$$\|v(t) - \bar{v}\|_{H^4}^2 + \|v_t(t)\|_{H^3}^2 + \|v_{tt}(t)\|_{H^1}^2 + \|u(t)\|_{H^4}^2 + \|u_t(t)\|_{H^2}^2 + \|u_{tt}(t)\|^2$$
$$+ \|\vec{w}(t)\|_{H^4}^2 + \|\vec{w}_t(t)\|_{H^2}^2 + \|\vec{w}_{tt}(t)\|^2 + \|\theta(t) - \bar{\theta}\|_{H^4}^2 + \|\theta_t(t)\|_{H^2}^2 + \|\theta_{tt}(t)\|^2$$
$$+ \int_0^t \Big(\|v - \bar{v}\|_{H^4}^2 + \|u\|_{H^5}^2 + \|u_t\|_{H^3}^2 + \|u_{tt}\|_{H^1}^2 + \|\vec{w}\|_{H^5}^2 + \|\vec{w}_t\|_{H^3}^2 + \|\vec{w}_{tt}\|^2$$
$$+ \|\theta - \bar{\theta}\|_{H^5}^2 + \|\theta_t\|_{H^3}^2 + \|\theta_{tt}\|_{H^1}^2 \Big)(\tau) \, d\tau \leq C_4, \tag{2.4.94}$$

$$\int_0^t \left(\|v_t\|_{H^4}^2 + \|v_{tt}\|_{H^2}^2 + \|v_{ttt}\|^2 \right)(\tau) \, d\tau \leq C_4. \tag{2.4.95}$$

Proof. Exploiting (2.1.16) and Lemmas 2.4.1–2.4.3, we easily obtain Lemma 2.4.4. The proof is complete. $\qquad\square$

Lemma 2.4.5. *Under assumptions of Theorem 2.1.3, for any* $(v_0, u_0, \vec{w}_0, \theta_0) \in H^4_+$ *, there exists a positive constant* $\gamma^{(1)}_4 = \gamma^{(1)}_4(C_4) \leq \gamma_2(C_2)$ *such that for any fixed* $\gamma \in (0, \gamma^{(1)}_4]$*, the following estimates hold for any* $t > 0$*, and* $\varepsilon \in (0, 1)$ *small enough,*

$$e^{\gamma t}\|u_{tt}(t)\|^2 + \int_0^t e^{\gamma \tau}\|u_{tty}(\tau)\|^2 \, d\tau \leq C_4 + C_4 \int_0^t e^{\gamma \tau}\|\theta_{tyy}(\tau)\|^2 \, d\tau, \qquad (2.4.96)$$

$$e^{\gamma t}\|\vec{w}_{tt}(t)\|^2 + \int_0^t e^{\gamma \tau}\|\vec{w}_{tty}(\tau)\|^2 \, d\tau \leq C_4 + C_4 \int_0^t e^{\gamma \tau}\|\vec{w}_{tyy}(\tau)\|^2 \, d\tau, \qquad (2.4.97)$$

$$e^{\gamma t}\|\theta_{tt}(t)\|^2 + \int_0^t e^{\gamma \tau}\|\theta_{tty}(\tau)\|^2 \, d\tau \leq C_4\varepsilon^{-3} + C_2\varepsilon^{-1}\int_0^t e^{\gamma \tau}\|\theta_{tyy}(\tau)\|^2 \, d\tau$$

$$+ C_1\varepsilon \int_0^t e^{\gamma \tau} \left(\|u_{tty}\|^2 + \|u_{tyy}\|^2 + \|\vec{w}_{tty}\|^2 + \|\vec{w}_{tyy}\|^2 \right)(\tau) \, d\tau.$$

$$(2.4.98)$$

Proof. Multiplying (2.4.27) by $e^{\gamma t}$, we have

$$\frac{1}{2}\frac{d}{dt}\left(e^{\gamma t}\|u_{tt}(t)\|^2\right) \leq \frac{\gamma}{2}e^{\gamma t}\|u_{tt}(t)\|^2 - C_1^{-1}e^{\gamma t}\|u_{tty}(t)\|^2 + C_2 e^{\gamma t}\Big(\|u_y(t)\|^2_{H^2}$$

$$+ \|\theta_t(t)\|^2 + \|\theta_{ty}(t)\|^2 + \|u_{ty}(t)\|^2 + \|v_y(t)\|^2 + \|\theta_{tt}(t)\|^2\Big).$$

$$(2.4.99)$$

Using (2.1.12) and Poincaré's inequality, we can derive

$$\|u_{tt}(t)\| \leq C_1\|u_{tty}(t)\|.$$

Exploiting (2.4.25) and Lemmas 2.2.3–2.2.4, we arrive at

$$e^{\gamma t}\|u_{tt}(t)\|^2 \leq C_4 - (C_1^{-1} - C_1\gamma)\int_0^t e^{\gamma \tau}\|u_{tty}(\tau)\|^2 \, d\tau + C_2 \int_0^t e^{\gamma \tau}\|\theta_{tt}(\tau)\|^2 \, d\tau$$

$$\leq C_4 - (C_1^{-1} - C_1\gamma)\int_0^t e^{\gamma \tau}\|u_{tty}(\tau)\|^2 \, d\tau + C_4 \int_0^t e^{\gamma \tau}\|\theta_{tyy}(\tau)\|^2 \, d\tau$$

which gives (2.4.96) if we take $\gamma > 0$ so small that $0 < \gamma \leq \min\left[\frac{1}{4C_1^2}, \gamma_2(C_2)\right]$. Multiplying (2.4.28) by $e^{\gamma t}$, using (2.4.24), Theorems 2.1.1–2.1.3 and Poincaré's inequality, we have

$$e^{\gamma t}\|\vec{w}_{tt}(t)\|^2 \leq C_4 - (C_1^{-1} - C_1\gamma)\int_0^t e^{\gamma \tau}\|\vec{w}_{tty}(\tau)\|^2 \, d\tau + C_4 \int_0^t e^{\gamma \tau}\|\vec{w}_{tyy}(\tau)\|^2 \, d\tau$$

which gives (2.4.97) if we take $\gamma > 0$ so small that $0 < \gamma \leq \min\left[\frac{1}{4C_1^2}, \gamma_2(C_2)\right]$.

Multiplying (2.4.29) by $e^{\gamma\tau}$, using (2.4.30)–(2.4.37), (2.4.25), Theorems 2.1.1–2.1.3 and Poincaré's inequality, for any $\varepsilon \in (0,1)$ small enough, we have

$$\frac{1}{2}e^{\gamma t}\|\sqrt{e_\theta}\theta_{tt}\|^2 \le C_4 + \frac{\gamma}{2}\int_0^t e^{\gamma\tau}\|\sqrt{e_\theta}\theta_{tt}\|^2(\tau)\,d\tau$$

$$+ \int_0^t e^{\gamma\tau}(A_1 + A_2 + A_3 + A_4 + A_5 + A_6 + A_7)(\tau)\,d\tau$$

$$\le C_4\varepsilon^{-3} + \frac{\gamma}{2}\int_0^t e^{\gamma\tau}\|\sqrt{e_\theta}\theta_{tt}(\tau)\|^2\,d\tau - (2C_1)^{-1}\int_0^t e^{\gamma\tau}\|\theta_{tty}(\tau)\|^2\,d\tau$$

$$+ C_2\int_0^t \left[e^{\gamma\tau}\|\theta_{tyy}(\tau)\|^2 + 2\varepsilon e^{\gamma\tau}\|\theta_{tty(\tau)}\|^2\right]\,d\tau$$

$$+ C_2\varepsilon^{-1}\int_0^t e^{\gamma\tau}\|\theta_{tt}(\tau)\|^2\,d\tau + \sup_{0\le\tau\le t}\|\theta_{tt}(\tau)\|\left(\int_0^t e^{\gamma\tau}\|u_{ty}(\tau)\|^2\,d\tau\right)^{1/4}$$

$$\times \left(\int_0^t e^{\gamma\tau}\|u_{tyy}(\tau)\|^2\,d\tau\right)^{1/4}\left[\int_0^t e^{\gamma\tau}(\|u_y\|_{H^1}^2 + \|\theta_y\|_{H^1}^2 + \|u_{ty}\|^2)(\tau)\,d\tau\right]^{1/2}$$

$$+ \varepsilon\int_0^t e^{\gamma\tau}(\|u_{tty}\|^2 + \|\vec{w}_{tty}\|^2 + \|\vec{w}_{tyy}\|^2)(\tau)\,d\tau$$

$$\le C_4\varepsilon^{-3} - \left(\frac{1}{2C_1} - 2\varepsilon - C_1\gamma\right)\int_0^t e^{\gamma\tau}\|\theta_{tty}(\tau)\|^2\,d\tau + C_2\varepsilon^{-1}\int_0^t e^{\gamma\tau}\|\theta_{tyy}(\tau)\|^2\,d\tau$$

$$+ \varepsilon\int_0^t e^{\gamma\tau}\left(\|u_{tty}\|^2 + \|u_{tyy}\|^2 + \|\vec{w}_{tty}\|^2 + \|\vec{w}_{tyy}\|^2\right)(\tau)\,d\tau + \varepsilon\sup_{0\le\tau\le t}\|\theta_{tt}(\tau)\|^2$$

which gives (2.4.98) if we take $\gamma > 0$ small enough. $\square$

Lemma 2.4.6. *Under assumptions of Theorem 2.1.3, for any* $(v_0, u_0, \vec{w}_0, \theta_0) \in H_+^4$, *there exists a positive constant* $\gamma_4^{(2)} \le \gamma_4^{(1)}$ *such that for any fixed* $\gamma \in (0, \gamma_4^{(2)}]$, *the following estimates hold for any* $t > 0$, *and* $\varepsilon \in (0,1)$ *small enough,*

$$e^{\gamma t}\|u_{ty}(t)\|^2 + \int_0^t e^{\gamma\tau}\|u_{tyy}(\tau)\|^2\,d\tau$$

$$\le C_2\varepsilon^{-6} + C_1\varepsilon^2\int_0^t e^{\gamma\tau}\left(\|u_{tty}\|^2 + \|\theta_{tyy}\|^2\right)(\tau)\,d\tau, \qquad (2.4.100)$$

$$e^{\gamma t}\|\vec{w}_{ty}(t)\|^2 + \int_0^t e^{\gamma\tau}\|\vec{w}_{tyy}(\tau)\|^2\,d\tau$$

$$\le C_2\varepsilon^{-6} + C_1\varepsilon^2\int_0^t e^{\gamma\tau}\|\vec{w}_{tty}(\tau)\|^2\,d\tau, \qquad (2.4.101)$$

$$e^{\gamma t}\|\theta_{ty}(t)\|^2 + \int_0^t e^{\gamma\tau}\|\theta_{tyy}(\tau)\|^2\,d\tau$$

$$\leq C_2\varepsilon^{-6} + C_1\varepsilon^2 \int_0^t e^{\gamma\tau}\left(\|u_{tyy}\|^2 + \|\theta_{tty}\|^2 + \|\vec{w}_{tyy}\|^2\right)(\tau)\,d\tau, \qquad (2.4.102)$$

$$e^{\gamma t}\left(\|u_{ty}(t)\|^2 + \|\vec{w}_{ty}(t)\|^2 + \|\theta_{ty}(t)\|^2\right)$$

$$+ \int_0^t e^{\gamma\tau}\left(\|u_{tyy}\|^2 + \|\vec{w}_{tyy}\|^2 + \|\theta_{tyy}\|^2\right)(\tau)\,d\tau$$

$$\leq C_2\varepsilon^{-6} + C_1\varepsilon^2 \int_0^t e^{\gamma\tau}(\|u_{tty}\|^2 + \|\vec{w}_{tty}\|^2 + \|\theta_{tty}\|^2)(\tau)\,d\tau. \qquad (2.4.103)$$

Proof. Adding up (2.4.100), (2.4.101) and (2.4.102), picking $\varepsilon \in (0,1)$ small enough, we obtain (2.4.103).

Multiplying (2.4.42) by $e^{\gamma t}$, using (2.4.47)–(2.4.48) and Lemmas 2.2.3–2.2.4, for any $\varepsilon \in (0,1)$ small enough, we have

$$\frac{1}{2}\frac{d}{dt}\left(e^{\gamma t}\|u_{ty}\|^2\right)$$

$$\leq \frac{1}{2}\gamma e^{\gamma t}\|u_{ty}\|^2 + e^{\gamma t}\Big[\varepsilon^2(\|u_{tyy}\|^2 + \|\theta_{tyy}\|^2 + \|u_{tyyy}\|^2)$$

$$+ C_2\varepsilon^{-6}\left(\|u_y\|_{H^2}^2 + \|\theta_t\|^2 + \|\theta_{ty}\|^2 + \|u_{ty}\|^2\right)\Big] - (2C_1)^{-1}e^{\gamma t}\|u_{tyy}\|^2$$

$$+ C_2 e^{\gamma t}\left(\|u_y\|_{H^1}^2 + \|\theta_t\|_{H^1}^2 + \|u_{ty}\|^2 + \|v_y\|_{H^1}^2\right). \qquad (2.4.104)$$

Integrating (2.4.104) with respect to t and using Poincaré's inequality, we have

$$e^{\gamma t}\|u_{ty}(t)\|^2 \leq C_2\varepsilon^{-6} - [(2C_1)^{-1} - 2\varepsilon^2 - C_1\gamma]\int_0^t e^{\gamma\tau}\|u_{tyy}(\tau)\|^2\,d\tau$$

$$+ C_1\varepsilon^2 \int_0^t e^{\gamma\tau}(\|u_{tty}\|^2 + \|\theta_{tyy}\|^2)(\tau)\,d\tau$$

which gives (2.4.100) if we take $\gamma > 0$ and $\varepsilon \in (0,1)$ so small that $0 < \varepsilon < \min[1, \frac{1}{8C_1}]$ and $0 < \gamma < \min[\gamma_4^{(1)}, \frac{1}{8C_1{}^2}] \equiv \gamma_4^{(2)}$.

Similarly, multiplying (2.4.51) by $e^{\gamma t}$, using (2.4.52)–(2.4.53), (2.4.55) and Theorems 2.1.1–2.1.3, for any $\varepsilon \in (0,1)$ small enough , we obtain

$$e^{\gamma t}\|\vec{w}_{ty}(t)\|^2 \leq C_2\varepsilon^{-6} - [(2C_1)^{-1} - C_1\varepsilon^2]\int_0^t e^{\gamma\tau}\|\vec{w}_{tyy}(\tau)\|^2\,d\tau$$

$$+ C_1\varepsilon^2 \int_0^t e^{\gamma\tau}\|\vec{w}_{tyyy}(\tau)\|^2\,d\tau$$

which gives (2.4.101) if we take $\varepsilon \in (0,1)$ small enough.

Multiplying (2.4.56) by $e^{\gamma t}$, using (2.4.57)–(2.4.61), (2.4.63) and Theorems 2.1.1–2.1.3, we derive

$$
\begin{aligned}
\frac{1}{2}\frac{d}{dt}\left(e^{\gamma t}\|\sqrt{e_\theta}\theta_{ty}\|^2\right) &\leq \frac{1}{2}\gamma e^{\gamma t}\|\sqrt{e_\theta}\theta_{ty}\|^2 + \varepsilon^2 e^{\gamma t}(\|\theta_{tyy}\|^2 + \|\theta_{tyyy}\|^2) \\
&+ C_2\varepsilon^{-6}e^{\gamma t}\left(\|\theta_{ty}\|^2 + \|\theta_y\|^2_{H^2} + \|u_y\|^2_{H^2} + \|v_y\|^2_{H^1}\right) - (2C_1)^{-1}e^{\gamma t}\|\theta_{tyy}\|^2 \\
&+ C_2\varepsilon^{-2}e^{\gamma t}\Big(\|u_y\|^2_{H^2} + \|\theta_t\|^2_{H^1} + \|u_{ty}\|^2 + \|\theta_y\|^2_{H^2} + \|v_y\|^2 \\
&+ \|\vec{w}_y\|^2_{H^2} + \|\vec{w}_{ty}\|^2\Big) + \varepsilon^2 e^{\gamma t}\|\theta_{tyy}\|^2 + C_2\varepsilon^2 e^{\gamma t}\|v_{ty}\|^2. \qquad (2.4.105)
\end{aligned}
$$

Integrating (2.4.105) with respect to t, we can deduce

$$
\begin{aligned}
e^{\gamma t}\|\theta_{ty}(t)\|^2 &\leq \varepsilon^2\int_0^t e^{\gamma\tau}\left(\|\vec{w}_{tyy}\|^2 + \|\theta_{tyyy}\|^2\right)(\tau)\,d\tau \\
&+ C_2\varepsilon^{-6} - \left((2C_1)^{-1} - \varepsilon^2\right)\int_0^t e^{\gamma\tau}\|\theta_{tyy}(\tau)\|^2\,d\tau
\end{aligned}
$$

which gives (2.4.102) for $\varepsilon \in (0,1)$ small enough.   □

Lemma 2.4.7. *Under assumptions of Theorem 2.1.3, for any $(v_0, u_0, \vec{w}_0, \theta_0) \in H^4_+$, there exists a positive constant $\gamma_4 \leq \gamma_4^{(2)}$ such that for any fixed $\gamma \in (0, \gamma_4]$, the following estimates hold for any $t > 0$, and $\varepsilon \in (0,1)$,*

$$
\begin{aligned}
&e^{\gamma t}(\|u_{tt}(t)\|^2 + \|\theta_{tt}(t)\|^2 + \|\vec{w}_{tt}(t)\|^2 + \|u_{ty}(t)\|^2 + \|\vec{w}_{ty}(t)\|^2 + \|\theta_{ty}(t)\|^2) \\
&+ \int_0^t e^{\gamma\tau}\left(\|u_{tty}\|^2 + \|u_{tyy}\|^2 + \|\theta_{tty}\|^2 + \|\theta_{tyy}\|^2 + \|\vec{w}_{tty}\|^2 + \|\vec{w}_{tyy}\|^2\right)(\tau)\,d\tau \\
&\leq C_4, \qquad (2.4.106)
\end{aligned}
$$

$$
\begin{aligned}
&e^{\gamma t}(\|v_{yyy}(t)\|^2_{H^1} + \|v_{yy}(t)\|^2_{W^{1,\infty}}) + \int_0^t e^{\gamma\tau}(\|v_{yyy}\|^2_{H^1} + \|v_{yy}\|^2_{W^{1,\infty}})(\tau)\,d\tau \\
&\leq C_4, \qquad (2.4.107)
\end{aligned}
$$

$$
\begin{aligned}
&e^{\gamma t}(\|u_{yyy}(t)\|^2_{H^1} + \|u_{yy}(t)\|^2_{W^{1,\infty}} + \|\vec{w}_{yyy}(t)\|^2_{H^1} + \|\vec{w}_{yy}(t)\|^2_{W^{1,\infty}} + \|\theta_{yyy}(t)\|^2_{H^1} \\
&+ \|\theta_{yy}(t)\|^2_{W^{1,\infty}} + \|v_{tyyy}(t)\|^2 + \|u_{tyy}(t)\|^2 + \|\vec{w}_{tyy}(t)\|^2 + \|\theta_{tyy}(t)\|^2) \\
&+ \int_0^t e^{\gamma\tau}\Big(\|u_{tt}\|^2 + \|\theta_{tt}\|^2 + \|\vec{w}_{tt}\|^2 + \|u_{yyyy}\|^2_{H^1} + \|u_{tyy}\|^2_{H^1} + \|\vec{w}_{yyyy}\|^2_{H^1} \\
&+ \|\vec{w}_{tyy}\|^2_{H^1} + \|\theta_{yyyy}\|^2_{H^1} + \|\theta_{tyy}\|^2_{H^1} + \|u_{yy}\|^2_{W^{2,\infty}} + \|u_{ty}\|^2_{W^{2,\infty}} \\
&+ \|\vec{w}_{yy}\|^2_{W^{2,\infty}} + \|\vec{w}_{ty}\|^2_{W^{2,\infty}} + \|\theta_{yy}\|^2_{W^{2,\infty}} + \|\theta_{ty}\|^2_{W^{2,\infty}} + \|v_{tyyy}\|^2_{H^1}\Big)(\tau)\,d\tau \\
&\leq C_4. \qquad (2.4.108)
\end{aligned}
$$

Proof. $(2.4.96) \times \varepsilon + (2.4.97) \times \varepsilon + (2.4.98) \times \varepsilon^{\frac{3}{2}} + (2.4.103)$, and taking $\varepsilon \in (0,1)$ small enough, we derive (2.4.106). Multiplying (2.4.72) by $e^{\gamma t}$, we can calculate

$$\frac{d}{dt}\left(e^{\gamma t}\left\|\frac{v_{yyyy}}{v}\right\|^2\right) + \left(C_1^{-1} - \gamma\right)e^{\gamma t}\left\|\frac{v_{yyy}}{v}\right\|^2 \le C_1 e^{\gamma t}\|E_1(t)\|^2. \qquad (2.4.109)$$

Choose $\gamma > 0$ so small that $0 < \gamma < \gamma_4 \equiv \min\left[\frac{1}{2C_1}, \gamma_4^{(2)}\right]$.

Integrating (2.4.109) with respect to t, we have

$$e^{\gamma t}\left\|\frac{v_{yyy}}{v}(t)\right\|^2 + \frac{1}{2C_1}\int_0^t e^{\gamma \tau}\left\|\frac{v_{yyy}}{v}(\tau)\right\|^2 d\tau \le C_3 + C_1\int_0^t e^{\gamma \tau}\|E_1(\tau)\|^2 d\tau, \quad \forall t > 0.$$

Using (2.4.106), we have

$$e^{\gamma t}\|v_{yyy}(t)\|^2 + \int_0^t e^{\gamma \tau}\|v_{yyy}(\tau)\|^2 d\tau \le C_4, \qquad \forall t > 0. \qquad (2.4.110)$$

By (2.4.9), (2.4.15), (2.4.20), (2.4.106) and Theorems 2.1.1–2.1.3, we have

$$e^{\gamma t}(\|u_{yyy}(t)\|^2 + \|\theta_{yyy}(t)\|^2 + \|\vec{w}_{yyy}(t)\|^2) \le C_4. \qquad (2.4.111)$$

By (2.4.12), (2.4.17), (2.4.22), (2.4.106) and Theorems 2.1.1–2.1.3, we get

$$\int_0^t e^{\gamma \tau}\left(\|u_{yyy}\|_{H^1}^2 + \|\theta_{yyy}\|_{H^1}^2 + \|\vec{w}_{yyy}\|_{H^1}^2\right)(\tau)\,d\tau \le C_4. \qquad (2.4.112)$$

Using the embedding theorem, the interpolation inequality and (2.4.109)–(2.4.112), we conclude

$$e^{\gamma t}\Big(\|u_{yyy}(t)\|^2 + \|\theta_{yyy}(t)\|^2 + \|\vec{w}_{yyy}(t)\|^2$$
$$+ \|u_{yy}(t)\|_{L^\infty}^2 + \|\theta_{yy}(t)\|_{L^\infty}^2 + \|\vec{w}_{yy}(t)\|_{L^\infty}^2\Big)$$
$$+ \int_0^t e^{\gamma \tau}\left(\|u_{yyy}\|_{H^1}^2 + \|\theta_{yyy}\|_{H^1}^2 + \|\vec{w}_{yyy}\|_{H^1}^2\right)(\tau)\,d\tau \le C_4. \qquad (2.4.113)$$

Using (2.4.77)–(2.4.79), (2.4.106) and Theorems 2.1.1–2.1.3, we derive

$$e^{\gamma t}(\|u_{tyy}(t)\|^2 + \|\theta_{tyy}(t)\|^2 + \|\vec{w}_{tyy}(t)\|^2) \le e^{\gamma t}\left(\|u_{tt}\|^2 + \|\vec{w}_{tt}\|^2 + \|\theta_{tt}\|^2\right)$$
$$+ e^{\gamma t}\left(\|u_y\|_{H^1}^2 + \|v_y\|_{H^1}^2 + \|\vec{w}_y\|_{H^1}^2 + \|\theta_t\|^2 + \|\theta_{ty}\|^2 + \|u_{ty}\|^2 + \|\vec{w}_{ty}\|^2\right) \le C_4.$$
$$\qquad (2.4.114)$$

By (2.4.12), (2.4.17), (2.4.22), (2.4.111)–(2.4.114), we conclude

$$e^{\gamma t}\Big(\|u_{yyy}(t)\|_{H^1}^2 + \|\theta_{yyy}(t)\|_{H^1}^2 + \|\vec{w}_{yyy}(t)\|_{H^1}^2 + \|u_{yy}(t)\|_{W^{1,\infty}}^2 + \|\vec{w}_{yy}(t)\|_{W^{1,\infty}}^2$$
$$+ \|\theta_{yy}(t)\|_{W^{1,\infty}}^2 + \|u_{tyy}(t)\|^2 + \|\vec{w}_{tyy}(t)\|^2 + \|\theta_{tyy}(t)\|^2\Big)$$
$$+ \int_0^t e^{\gamma \tau}\Big(\|u_{yyy}\|_{H^1}^2 + \|\theta_{yyy}\|_{H^1}^2 + \|\vec{w}_{yyy}\|_{H^1}^2 + \|u_{yy}\|_{W^{1,\infty}}^2 + \|\vec{w}_{yy}\|_{W^{1,\infty}}^2$$
$$+ \|\theta_{yy}\|_{W^{1,\infty}}^2 + \|u_{tyy}\|^2 + \|\vec{w}_{tyy}\|^2 + \|\theta_{tyy}\|^2\Big)(\tau)\,d\tau \le C_4, \quad \forall t > 0. \qquad (2.4.115)$$

Using (2.4.23)–(2.4.24), (2.4.26), (2.4.115) and Theorem 2.1.3, we can deduce

$$\int_0^t e^{\gamma\tau}\left(\|u_{tt}\|^2 + \|\vec{w}_{tt}\|^2 + \|\theta_{tt}\|^2\right)(\tau)\,d\tau \leq C_4, \quad \forall t > 0. \tag{2.4.116}$$

Analogously, by (2.4.49), (2.4.55), (2.4.63), (2.4.106), (2.4.115) and Theorem 2.1.3, we derive

$$\int_0^t e^{\gamma\tau}(\|u_{tyyy}\|^2 + \|\vec{w}_{tyyy}\|^2 + \|\theta_{tyyy}\|^2)(\tau)\,d\tau \leq C_4, \quad \forall t > 0. \tag{2.4.117}$$

Multiplying (2.4.87) by $e^{\gamma t}$, using (2.4.83), (2.4.106), (2.4.110)–(2.4.117) and Theorem 2.1.3, for any fixed $\gamma \in (0, \gamma_4)$, we have

$$e^{\gamma t}\left\|\frac{v_{yyyy}}{v}\right\|^2 + \frac{1}{2C_1}\int_0^t e^{\gamma\tau}\left\|\frac{v_{yyyy}}{v}\right\|^2(\tau)\,d\tau \leq C_4 + C_1\int_0^t e^{\gamma\tau}\|E_2(\tau)\|^2\,d\tau$$

$$\leq C_4, \quad \forall t > 0.$$

Thus

$$e^{\gamma t}\|v_{yyyy}(t)\|^2 + \int_0^t e^{\gamma\tau}\|v_{yyyy}(\tau)\|^2\,d\tau \leq C_4, \quad \forall t > 0 \tag{2.4.118}$$

which, combined with (2.4.110), (2.4.118) and the embedding theorem, gives (2.4.107).

On the other hand, by (2.4.89)–(2.4.91), (2.4.115), (2.4.117) and Theorem 2.1.3, we have

$$\int_0^t e^{\gamma\tau}\Big(\|u_{yyyyy}\|^2 + \|\theta_{yyyyy}\|^2 + \|\vec{w}_{yyyyy}\|^2 + \|u_{yy}\|^2_{W^{2,\infty}}$$

$$+ \|\vec{w}_{yy}\|^2_{W^{2,\infty}} + \|\theta_{yy}\|^2_{W^{2,\infty}}\Big)(\tau)\,d\tau \leq C_4 \tag{2.4.119}$$

which, combined with (2.4.115)–(2.4.119), gives (2.4.108). □

Lemma 2.4.8. *Under assumptions of Theorem 2.1.3, for any $(v_0, u_0, \vec{w}_0, \theta_0) \in H^4_+$, for any fixed $\gamma \in (0, \gamma_4]$, the following estimates hold for any $t > 0$,*

$$e^{\gamma t}\Big(\|v(t) - \bar{v}\|^2_{H^4} + \|u(t)\|^2_{H^4} + \|\theta(t) - \bar{\theta}\|^2_{H^4} + \|v_t\|^2_{H^3}$$

$$+ \|v_{tt}(t)\|^2_{H^1} + \|u_t(t)\|^2_{H^2} + \|u_{tt}(t)\|^2 + \|\theta_t(t)\|^2_{H^2}$$

$$+ \|\theta_{tt}(t)\|^2 + \|\vec{w}_t(t)\|^2_{H^3} + \|\vec{w}_t(t)\|^2_{H^2} + \|\vec{w}_{tt}(t)\|^2\Big) \leq C_4, \tag{2.4.120}$$

$$\int_0^t e^{\gamma\tau}\Big(\|v(t) - \bar{v}\|^2_{H^4} + \|u\|^2_{H^5} + \|\theta(t) - \bar{\theta}\|^2_{H^5} + \|\vec{w}\|^2_{H^5} + \|u_t\|^2_{H^3}$$

$$+ \|u_{tt}\|^2_{H^1} + \|\vec{w}_t\|^2_{H^3} + \|\vec{w}_{tt}\|^2_{H^1} + \|\theta_t\|^2_{H^3} + \|\theta_{tt}\|^2_{H^1}$$

$$+ \|v_t\|^2_{H^4} + \|v_{tt}\|^2_{H^2} + \|v_{ttt}\|^2\Big)(\tau)\,d\tau \leq C_4. \tag{2.4.121}$$

Proof. Using (2.1.16), Theorems 2.1.1–2.1.2 and Lemmas 2.4.1–2.4.7, we can prove the conclusion.   □

Proof of Theorem 2.1.3. Exploiting Lemmas 2.4.1–2.4.8, we complete the proof of Theorem 2.1.3.   □

2.5 Bibliographic Comments

In this section, we shall recall some related known results in this direction.

For the case $\vec{w} = 0$, there are many results on the global existence and asymptotic behavior of solutions to problem (2.1.16)–(2.1.17), (2.1.19)–(2.1.22) with different constitutive assumptions; we refer to Jiang [26, 27, 28, 29], Kawashima and Nishida [31], Kawohl [32], Kazhikhov [34], Kazhikhov and Shelukhin [36], Okada and Kawashima [47], Qin [49, 50, 52, 55, 54], Qin and Muñoz Rivera [70], and Wang [76]. Among these cases we would like to mention two classes of models: an ideal gas and a real viscous gas. For the former case, i.e., for the case of $\vec{w} = 0$ and an ideal gas whose constitutive relations take the following form,

$$e = C_V \theta, \quad E = C_V \theta + \frac{1}{2}(u^2 + |\vec{w}|^2), \quad p = R\frac{\theta}{v}, \qquad (2.5.122)$$

with suitable positive constants C_V, R, the global existence and asymptotic behavior of smooth (generalized) solutions to the system (2.1.16), (2.1.17), (2.1.19) were established by many authors; we refer to Jiang [28, 29], Kawashima and Nishida [31], Kazhikhov [34], Kazhikhov and Shelukhin [36], Okada and Kawashima [47], Qin [49, 50, 51, 52, 55, 54, 56, 57], Qin, Wu and Liu [73] on the initial boundary value problems and the Cauchy problem. In detail, Qin [49, 50] established the existence and asymptotic behavior solutions in H^1 to (2.1.16), (2.1.17), (2.1.19)–(2.1.21) for a viscous ideal gas (2.5.122) in bounded domain in $\mathbb{R}$, for which Zheng and Qin [83] obtained the existence of maximal attractors (see also for a viscous ideal gas (2.5.122) in bounded annular domains $G_n = \{x \in \mathbb{R}^n | 0 < a < |x| < b\}$ $(n = 2, 3)$ in $\mathbb{R}^n$ for a viscous spherically symmetric ideal gas). For the latter case, i.e., for the case of $\vec{w} = 0$ and a real gas with the same assumptions as those in (2.1.25)–(2.1.36), Qin [55] (see also, Qin [51, 52, 54] with some stronger growth assumptions) established the existence and exponential stability of a C_0-semigroup generated by the solutions to (2.1.16), (2.1.17), (2.1.19)–(2.1.21) in the subspace of $H^i \times H^i \times H^i$ $(i = 1, 2)$ for a viscous ideal gas (2.5.122) in a bounded domain in $\mathbb{R}$.

For the case of $\vec{w} \neq 0$, an ideal flow (2.5.122) which is the special case of $q = r = 0$ of the problem (2.1.16)–(2.1.21), Qin [70] proved the exponential stability and existence of attractors; Wang [76] investigated the global existence, uniqueness, regularity in H^1. In this chapter, under more general assumptions (2.1.25)–(2.1.36) on the constitutive relations than those in [73], we establish the global existence uniqueness and asymptotic behavior of solutions in H^1 and H^2.

The novelties of this chapter consist of the following aspects: (1) the more general constitutive relations and growth assumptions (2.1.25)–(2.1.36) are studied, the related results in H^1 in this chapter have improved and extended those in [73]; (2) the global existence and exponential stability of solutions in H^1 and H^2 are established for the model under consideration; (3) the results in H^2 and H^4 are obtained first for the model under consideration.

Chapter 3

Regularity and Exponential Stability of the pth Power Newtonian Fluid in One Space Dimension

3.1 Introduction

In this chapter, we are interested in the regularity and exponential stability of solutions in H^i ($i = 2, 4$) for a pth power Newtonian fluid undergoing one-dimensional longitudinal motions. We assume that the pressure $\mathcal{P}$, in terms of the absolute temperature θ and the specific volume u, is given by

$$\mathcal{P} = \frac{\theta}{u^p} \tag{3.1.1}$$

with the pressure exponent $p \geq 1$.

The balance laws of mass, momentum, and energy in Lagrangian form are as follows:

$$u_t = v_x, \tag{3.1.2}$$

$$v_t = \left(-\mathcal{P} + \mu \frac{v_x}{u}\right)_x, \tag{3.1.3}$$

$$c_v \theta_t = \left(-\mathcal{P} + \mu \frac{v_x}{u}\right) v_x + \left(\kappa \frac{\theta_x}{u}\right)_x. \tag{3.1.4}$$

Here, u, v, θ are specific volume, velocity, and absolute temperature, respectively. The positive constants c_v, μ, κ represent specific heat, viscosity and conductivity, respectively. Since the magnitude of the specific heat c_v plays no role in the mathematical analysis of the system, in what follows we will assume the scaling

$$c_v = 1.$$

We consider a typical initial boundary value problem for (3.1.2)–(3.1.4) in the reference domain $\{(x,t) : 0 < x < 1, t \geq 0\}$ under the initial conditions

$$u(x,0) = u_0(x), \ v(x,0) = v_0(x), \ \theta(x,0) = \theta_0(x), \quad x \in [0,1] \tag{3.1.5}$$

and boundary conditions

$$v(0,t) = v(1,t) = 0, \ \theta_x(0,t) = \theta_x(1,t) = 0. \tag{3.1.6}$$

Obviously, when $p = 1$, (3.1.1) reduces to the case of a polytropic ideal gas (see (3.4.1)). This chapter mainly continues to the case of $p > 1$, which was selected from [61].

The notation in this chapter is standard. We put $\|\cdot\| = \|\cdot\|_{L^2[0,1]}$. Subscripts t and x denote the (partial) derivatives with respect to t and x, respectively. We use C_i $(i = 1, 2, 4)$ to denote a generic positive constant depending on the $H^i[0,1]$ norm of initial data (u_0, v_0, θ_0), $\min_{x \in [0,1]} u_0(x)$ and $\min_{x \in [0,1]} \theta_0(x)$, but independent of time variable t.

For convenience and without loss of generality, we may assume $\int_0^1 u_0(x)\,dx = 1$. Then from conservation of mass and boundary condition (3.1.6), we have

$$\int_0^1 u(x,t)\,dx = 1. \tag{3.1.7}$$

We define two spaces as

$$H_+^2 = \Big\{(u,v,\theta) \in H^2[0,1] \times H^2[0,1] \times H^2[0,1] : u(x) > 0, \ \theta(x) > 0,$$
$$\forall x \in [0,1], v(0) = v(1) = 0, \theta'(0) = \theta'(1) = 0\Big\}$$

and

$$H_+^4 = \Big\{(u,v,\theta) \in H^4[0,1] \times H^4[0,1] \times H^4[0,1] : u(x) > 0, \ \theta(x) > 0,$$
$$\forall x \in [0,1], v(0) = v(1) = 0, \theta'(0) = \theta'(1) = 0\Big\}.$$

Now our main results in this chapter read as follows.

Theorem 3.1.1. *Suppose that* $(u_0, v_0, \theta_0) \in H_+^2$ *and the compatibility conditions hold. Then there exists a unique generalized global solution* $(u(t), v(t), \theta(t)) \in H_+^2$ *to the problem* (3.1.2)–(3.1.6) *verifying that for any* $(x,t) \in [0,1] \times [0,+\infty)$,

$$0 < C_1^{-1} \leq u(x,t), \quad \theta(x,t) \leq C_1 \tag{3.1.8}$$

and for any $t > 0$,

$$\|u(t) - 1\|_{H^2}^2 + \|v(t)\|_{H^2}^2 + \|\theta(t) - \bar{\theta}\|_{H^2}^2 + \|v_t(t)\|^2 + \|\theta_t(t)\|^2 \tag{3.1.9}$$

$$+ \int_0^t \Big(\|u - 1\|_{H^2}^2 + \|v\|_{H^3}^2 + \|\theta - \bar{\theta}\|_{H^3}^2 + \|v_t\|_{H^1}^2 + \|\theta_t\|_{H^1}^2\Big)(\tau)\,d\tau \leq C_2.$$

Moreover, there exists a constant $\gamma_2 = \gamma_2(C_2) > 0$ such that for any fixed $\gamma \in (0, \gamma_2]$, the following inequality holds for any $t > 0$,

$$e^{\gamma t}\left(\|u(t) - 1\|_{H^2}^2 + \|v(t)\|_{H^2}^2 + \|\theta(t) - \overline{\theta}\|_{H^2}^2 + \|v_t(t)\|^2 + \|\theta_t(t)\|^2 \right) \qquad (3.1.10)$$

$$+ \int_0^t e^{\gamma \tau}\left(\|u - 1\|_{H^2}^2 + \|v\|_{H^3}^2 + \|\theta - \overline{\theta}\|_{H^3}^2 + \|v_t\|_{H^1}^2 + \|\theta_t\|_{H^1}^2 \right)(\tau)\, d\tau \leq C_2$$

where $\overline{\theta} = \int_0^1 \left(\theta_0 + \frac{1}{2}v_0^2\right) dx$.

Theorem 3.1.2. *Suppose that $(u_0, v_0, \theta_0) \in H_+^4$ and the compatibility conditions hold. Then there exists a unique global solution $(u(t), v(t), \theta(t)) \in H_+^4$ to the problem (3.1.2)–(3.1.6) such that for any $t > 0$,*

$$\|u(t) - 1\|_{H^4}^2 + \|v(t)\|_{H^4}^2 + \|\theta(t) - \overline{\theta}\|_{H^4}^2 + \|v_t(t)\|_{H^2}^2$$
$$+ \|v_{tt}(t)\|^2 + \|\theta_t(t)\|_{H^2}^2 + \|\theta_{tt}(t)\|^2$$
$$+ \int_0^t \left(\|u - 1\|_{H^4}^2 + \|v\|_{H^5}^2 + \|\theta - \overline{\theta}\|_{H^5}^2 + \|v_t\|_{H^3}^2 \right.$$
$$\left. + \|\theta_t\|_{H^3}^2 + \|v_{tt}\|_{H^1}^2 + \|\theta_{tt}\|_{H^1}^2 \right)(\tau)\, d\tau \leq C_4. \qquad (3.1.11)$$

Moreover, there exists a constant $\gamma_4 = \gamma_4(C_4) > 0$ such that for any fixed $\gamma \in (0, \gamma_4]$, the following inequality holds for any $t > 0$,

$$e^{\gamma t}\left(\|u(t) - 1\|_{H^4}^2 + \|v(t)\|_{H^4}^2 + \|\theta(t) - \overline{\theta}\|_{H^4}^2 + \|v_t(t)\|_{H^2}^2 \right.$$
$$\left. + \|v_{tt}(t)\|^2 + \|\theta_t(t)\|_{H^2}^2 + \|\theta_{tt}(t)\|^2 \right)$$
$$+ \int_0^t e^{\gamma \tau}(\|u - 1\|_{H^4}^2 + \|v\|_{H^5}^2 + \|\theta - \overline{\theta}\|_{H^5}^2 + \|v_t\|_{H^3}^2 + \|\theta_t\|_{H^3}^2$$
$$+ \|v_{tt}\|_{H^1}^2 + \|\theta_{tt}\|_{H^1}^2)(\tau)\, d\tau \leq C_4. \qquad (3.1.12)$$

Corollary 3.1.1. *Under assumptions of Theorem 3.1.2, (3.1.12) implies that $(u(t), v(t), \theta(t))$ is the classical solution verifying that for any fixed $\gamma \in (0, \gamma_4]$ and for any $t > 0$,*

$$\|(u(t) - 1, v(t), \theta(t) - \overline{\theta})\|_{(C^{3+1/2})^3}^2 \leq C_4 e^{-\gamma t}.$$

Remark 3.1.1. Obviously, it is easy to see that the similar results hold for the boundary conditions

$$v(0, t) = v(1, t) = 0, \ \theta(0, t) = \theta(1, t) = \text{const.} > 0.$$

Remark 3.1.2. Theorems 3.1.1–3.1.2 have improved the results in [37].

3.2   Proof of Theorem 3.1.1

In this section, we shall complete the proof of Theorem 3.1.1 and assume that the assumptions in Theorem 3.1.1 are valid. We begin with a technical lemma selected from [46].

Lemma 3.2.1. *Let $\lambda(t)$ (≥ 0) and $\omega(t)$ be continuous functions satisfying that there exist positive constants C_i $(i = 1, 2, 3, 4)$ such that*

$$C_1 e^{C_2(t-\tau)} \leq \exp\left\{\int_\tau^t \omega(s)ds\right\} \leq C_3 e^{C_4(t-\tau)}, \quad 0 \leq \tau \leq t. \tag{3.2.1}$$

We denote $\Lambda(t)$ by

$$\Lambda(t) = \int_t^{t+1} \lambda(\tau)d\tau.$$

Then

$$C^{-1} \lim_{t \to +\infty} \inf \Lambda(t) \leq \lim_{t \to +\infty} \inf \int_0^t \exp\left\{-\int_\tau^t \omega(s)ds\right\}\lambda(\tau)d\tau \tag{3.2.2}$$

$$\leq \lim_{t \to +\infty} \sup \int_0^t \exp\left\{-\int_\tau^t \omega(s)ds\right\}\lambda(\tau)d\tau \leq C \lim_{t \to +\infty} \sup \Lambda(t)$$

holds.

Proof. Only the estimate from above in (3.2.2) will be shown, the estimate from below is derived similarly. By use of (3.2.1),

$$\int_0^t \exp\left\{-\int_\tau^t \omega(s)ds\right\}\lambda(\tau)d\tau$$

$$\leq \exp\left\{-\int_0^t \omega(s)ds\right\}\int_0^{T+1} \exp\left\{\int_0^\tau \omega(s)ds\right\}\lambda(\tau)d\tau$$

$$+ \sum_{j=0}^{\max\{0,[t-T]-1\}} \int_{t-j-1}^{t-j} \exp\left\{-\int_0^\tau \omega(s)ds\right\}\lambda(\tau)d\tau$$

$$\leq C(T)\left\{-\int_0^t \omega(s)ds\right\} + C_1^{-1}\left(\sup_{t \geq T} \Lambda(t)\right)\sum_{j=0}^{\max\{0,[t-T]-1\}} e^{-C_2 j}$$

holds. Here $[\cdot]$ is a Gaussian symbol. Therefore the desired estimate easily follows. □

Lemma 3.2.2. *If $(u_0, v_0, \theta_0) \in H^1[0,1] \times H_0^1[0,1] \times H^1[0,1]$, then there exists a unique generalized global solution $(u(t), v(t), \theta(t)) \in H^1[0,1] \times H_0^1[0,1] \times H^1[0,1]$*

to the problem (3.1.2)–(3.1.6) satisfying

$$0 < C_1^{-1} \le u(x,t) \le C_1, \quad \forall (x,t) \in [0,1] \times [0,+\infty), \tag{3.2.3}$$

$$u_t, v_t, \theta_t, \theta_x, v_x, u_x, v_{xx}, \theta_{xx} \in L^2([0,+\infty), L^2[0,1]), \tag{3.2.4}$$

$$0 < \theta(x,t) \le C_1, \quad \forall (x,t) \in [0,1] \times [0,+\infty), \tag{3.2.5}$$

$$\|u(t)\|_{H^1}^2 + \|v(t)\|_{H^1}^2 + \|\theta(t)\|_{H^1}^2 + \int_0^t (\|u_x\|^2 + \|v\|_{H^2}^2$$
$$+ \|\theta_x\|_{H^1}^2 + \|v_t\|^2 + \|\theta_t\|^2)(\tau) \, d\tau \le C_1, \quad \forall t > 0, \tag{3.2.6}$$

and there exist positive constants λ, t_0, Λ, independent of t, such that as $t \ge t_0$,

$$\|u_x(t)\|^2 + \|v_x(t)\|^2 + \|\theta_x(t)\|^2 \le \Lambda e^{-\lambda t}, \tag{3.2.7}$$

$$\max_{x \in [0,1]} (|u(x,t) - 1| + |v(x,t)| + |\theta(x,t) - \bar\theta|) \le \Lambda e^{-\lambda t}. \tag{3.2.8}$$

Proof. The existence of a generalized global solution in $H^1[0,1]$ and estimates (3.2.4)–(3.2.8) were obtained by Lewicka and Watson [37]. For convenience of the reader, we give a simple proof. First, we note that the entropy η of a pth power Newtonian fluid is a concave function

$$\eta(u,\theta) = \log \theta + h(u), \tag{3.2.9}$$

where

$$h(u) = \begin{cases} \log u, & p = 1, \\ \frac{1}{p-1}(1 - u^{1-p}), & p > 1, \end{cases}$$

which satisfies the following standard entropy identity:

$$\eta_t = \mu \frac{v_x^2}{u\theta} + \kappa \frac{\theta_x^2}{u\theta^2} - \left(\frac{q}{\theta}\right)_x. \tag{3.2.10}$$

Set

$$e = c_v \theta, \quad \sigma = -\frac{\theta}{u^p} + \mu \frac{v_x}{u}, \quad q = -\kappa \frac{\theta_x}{u}.$$

By combining (3.1.2)–(3.1.4) and (3.2.10), we obtain the identity,

$$\left(e + \frac{1}{2}v^2 - \bar\theta \eta\right)_t = \left(\sigma v - \left(1 - \frac{\bar\theta}{\theta}\right) q\right)_x - \bar\theta \left(\mu \frac{v_x^2}{u\theta} + \kappa \frac{\theta_x^2}{u\theta^2}\right). \tag{3.2.11}$$

Integrating (3.2.9) with respect to x and then using Jensen's inequality and (3.1.7), we infer from the fact $\int_0^1 u\,dx = 1$,

$$\int_0^1 \eta \, dx \le \log\left(\int_0^1 \theta \, dx\right) + h\left(\int_0^1 u \, dx\right) = \log\left(\int_0^1 \theta \, dx\right). \tag{3.2.12}$$

Integrating (3.2.11) over $[0,1] \times [0,t]$, and noting the boundary conditions, we arrive at

$$\int_0^1 \left(\theta + \frac{1}{2} v^2 \right) dx + \bar\theta \int_0^t \int_0^1 \left(\mu \frac{v_x^2}{u\theta} + \kappa \frac{\theta_x^2}{u\theta^2} \right) dxds \le C_1 + \int_0^1 \bar\theta \eta dx$$

which, along with (3.2.12), gives

$$\lambda \le \int_0^1 \theta(x,t) dx \le \Lambda, \tag{3.2.13}$$

$$\int_0^1 v^2 dx + \bar\theta \int_0^t \int_0^1 \left(\mu \frac{v_x^2}{u\theta} + \kappa \frac{\theta_x^2}{u\theta^2} \right) dxds \le C_1. \tag{3.2.14}$$

Set

$$\theta_m = \max_{x \in [0,1]} \theta(x,t), \quad u_m = \max_{x \in [0,1]} u(x,t), \quad v_m = \max_{x \in [0,1]} v(x,t).$$

By a straightforward calculation, we have

$$\theta(x,t) \le \left[\theta^{1/2}(y,t) + \frac{1}{2} \int_0^1 \frac{|\theta_x|}{\theta^{1/2}} dx \right]^2 \le 2 \left[\theta(y,t) + \frac{1}{4} \left(\int_0^1 \frac{|\theta_x|}{\theta^{1/2}} dx \right)^2 \right]$$

$$\le 2 \left[\theta(y,t) + \frac{1}{4} \left(\int_0^1 u\theta dx \right) \left(\int_0^1 \frac{\theta_x^2}{u\theta^2} dx \right) \right]. \tag{3.2.15}$$

Integrating (3.2.15) with respect to y over $[0,1]$, by (3.2.13), we obtain

$$\theta_m(t) \le C_1 \left(1 + u_m(t) \int_0^1 \frac{\theta_x^2}{u\theta^2} dx \right). \tag{3.2.16}$$

In a similar manner, we have

$$\theta(x,t) \ge C_1^{-1} - C_1 \int_0^1 \frac{\theta_x^2}{\theta^2} dx. \tag{3.2.17}$$

To prove (3.2.3), we divide the proof into three steps.

Step 1. Integrating (3.1.3) over $[x,1] \times [\tau,t]$, we get

$$\int_1^x (v(r,t) - v(r,\tau)) \, dr = \int_\tau^t \sigma(x,s) ds - \int_\tau^t \sigma(1,s) ds$$

$$= - \int_\tau^t \sigma(1,s) ds - \int_\tau^t \frac{\theta}{u^p}(x,s) ds + \mu \ln u(x,s) \Big|_{s=\tau}^{s=t}.$$

Setting $M(x,\tau,t) := \int_1^x [v(r,t) - v(r,\tau)] dr$, and the impulse

$$I(\tau,t) := \int_\tau^t \sigma(1,s) ds,$$

we rewrite the above equation in the form

$$\int_\tau^t \frac{\theta}{u^p}(x,s)ds = \mu \log u(x,t) - \mu \log u(x,\tau) - I(\tau,t) - M(x,\tau,t). \qquad (3.2.18)$$

Multiplying (3.2.18) by p/μ and taking exponentials, we have readily that

$$\frac{d}{dt}\left[u^p \cdot \exp\left\{-\frac{p}{\mu}M - \frac{p}{\mu}I\right\}\right] = \frac{p}{\mu}\theta \exp\left\{-\frac{p}{\mu}M - \frac{p}{\mu}I\right\}.$$

Hence,

$$u^p(x,t) \cdot \exp\left\{-\frac{p}{\mu}M(x,\tau,t) - \frac{p}{\mu}I(\tau,t)\right\}$$

$$= u^p(x,\tau) + \int_\tau^t \frac{p}{\mu}\theta(x,s)\exp\left\{-\frac{p}{\mu}M(x,\tau,s) - \frac{p}{\mu}I(\tau,s)\right\}ds. \qquad (3.2.19)$$

By (3.2.14), we have

$$|M(x,\tau,t)|^2 \leq 2\int_0^1 \left[v^2(x,t) + v^2(x,\tau)\right]dx \leq C_1. \qquad (3.2.20)$$

Introducing (3.2.20) in (3.2.19), we arrive at

$$C_1\left[u^p(x,\tau) + \int_\tau^t \theta(x,s)e^{-\frac{p}{\mu}I(\tau,s)}ds\right] \leq u^p(x,t)e^{-\frac{p}{\mu}I(\tau,t)} \qquad (3.2.21)$$

$$\leq C_1\left[u^p(x,\tau) + \int_\tau^t \theta(x,s)e^{-\frac{p}{\mu}I(\tau,s)}ds\right].$$

Step 2. In this step, we shall prove the uniform upper bound on u. First, from (3.1.7) it follows that

$$\int_0^1 u^p(x,t)dx \leq u_m^{p-1}(t)\int_0^1 u(x,t)dx = u_m^{p-1}. \qquad (3.2.22)$$

On the other hand, by Jensen's inequality, we have

$$\int_0^1 u^p(x,t)dx \geq \left(\int_0^1 u(x,t)dx\right)^p = 1. \qquad (3.2.23)$$

From the right-hand inequality in (3.2.21) with $\tau = 0$, (3.2.16) shows that,

$$u_m^p(t) \cdot e^{-\frac{p}{\mu}I(0,t)}$$

$$\leq C_1\left[1 + \int_0^t e^{-\frac{p}{\mu}I(0,s)}ds + \int_0^t u_m(s)e^{-\frac{p}{\mu}I(0,s)}\left(\int_0^1 \frac{\theta_x^2}{u\theta^2}dx\right)(s)ds\right]. \qquad (3.2.24)$$

Since by (3.2.23) $u_m^p \geq 1$, it is clear that $u_m(t) \leq u_m^p(t)$, and thus from (3.2.24), by means of the Gronwall inequality, we obtain

$$u_m^p(t) \cdot e^{-\frac{p}{\mu}I(0,t)} \leq C_1 \left(1 + \int_0^t e^{-\frac{p}{\mu}I(0,s)}ds\right) \cdot \exp\left\{\int_0^t \int_0^1 \frac{\theta_x^2}{u\theta^2}dxds\right\}.$$

From (3.2.14) it follows that

$$u_m^p(t) \leq C_1 e^{\frac{p}{\mu}I(0,t)} \left(1 + \int_0^t e^{-\frac{p}{\mu}I(0,s)}ds\right). \tag{3.2.25}$$

On the other hand, setting $\tau = 0$ in the left-hand inequality (3.2.21), then integrating over the spatial interval $[0,1]$, and utilizing the estimates (3.2.22), (3.2.23) and (3.2.13), we have the following bound,

$$u_m^{p-1} \geq C_1^{-1} e^{\frac{p}{\mu}I(0,t)} \left(1 + \int_0^t e^{-\frac{p}{\mu}I(0,s)}ds\right). \tag{3.2.26}$$

Now, (3.2.25) and (3.2.26) give

$$u_m^p(t) \leq C_1 u_m^{p-1}(t),$$

from which we conclude the existence of a constant $C_1 > 0$ such that

$$u(x,t) \leq C_1. \tag{3.2.27}$$

Step 3. Our next concern will be the lower bound on u. Integrating (3.2.21) in x over $[0,1]$ and recalling (3.2.23), (3.2.27) and (3.2.13), we see that

$$C_1^{-1} \leq e^{\frac{p}{\mu}I(\tau,t)} \left(1 + \int_\tau^t e^{-\frac{p}{\mu}I(\tau,s)}ds\right) \leq C_1. \tag{3.2.28}$$

Setting $\tau = 0$ in the left-hand inequality in (3.2.21) while utilizing (3.2.17) and (3.2.28), we have

$$u^p(x,t) \geq C_1^{-1} e^{\frac{p}{\mu}I(0,t)} \cdot \left[u_0^p(x) + \int_0^t e^{-\frac{p}{\mu}I(0,s)}\theta(x,s)ds\right]$$

$$\geq C_1^{-1} e^{\frac{p}{\mu}I(0,t)} \cdot \left[u_0^p(x) + \int_0^t e^{-\frac{p}{\mu}I(0,s)}ds\right.$$

$$\left. -C_1 \int_0^t e^{-\frac{p}{\mu}I(0,s)} \left(\int_0^1 \frac{\theta_x^2}{\theta^2}dx\right)(s)ds\right]$$

$$\geq C_1^{-1} - C_1 \int_0^t e^{\frac{p}{\mu}I(s,t)} \left(\int_0^1 \frac{\theta_x^2}{\theta^2}dx\right)(s)ds. \tag{3.2.29}$$

Now, by Gronwall's inequality applied to (3.2.28), we obtain

$$C_1^{-1} e^{C_1^{-1}(t-\tau)} \leq e^{-\frac{p}{\mu}I(\tau,t)} \leq C_1 e^{C_1(t-\tau)}.$$

Thus, by virtue of Lemma 3.2.1 with $\omega(t) = -\sigma(1,t)$ and $\lambda(t) = \int_0^1 \frac{\theta_x^2}{\theta^2}(x,t)dx$, we conclude that

$$\lim_{t\to+\infty} \int_0^t e^{\frac{P}{\mu}I(s,t)} \left(\int_0^1 \frac{\theta_x^2}{\theta^2}dx \right)(s)ds \leq C_1 \lim_{t\to+\infty} \int_t^{t+1} \int_0^1 \frac{\theta_x^2}{\theta^2}dxds. \qquad (3.2.30)$$

On the other hand, (3.2.14) and (3.2.27) imply that the function $\int_0^1 \frac{\theta_x^2}{\theta^2}dx$ is integrable in $[0,+\infty)$, so the right-hand side of (3.2.30) equals zero. Now since the right-hand side of the first inequality in (3.2.29) is a continuous and positive function of t, in view of (3.2.30), it implies that

$$u(x,t) \geq C_1^{-1}$$

which, together with (3.2.27), gives (3.2.3).

The balance of momentum (3.1.3) can be rewritten in the form:

$$\left(\mu \frac{u_x}{u} - v \right)_t = \left(\frac{\theta}{u^p} \right)_x.$$

Multiplying by $\mu u_x/u - v$ and integrating over $[0,1] \times [0,t]$, using the interpolation inequality and the Young inequality, we derive

$$\int_0^1 u_x^2 dx + \int_0^t \int_0^1 (u_x^2 + \theta_x^2 + \theta^2 u_x^2)dxds \leq C_1. \qquad (3.2.31)$$

In a similar manner, multiplying (3.1.3) by v, integrating over $[0,1] \times [0,t]$ and using integrations by parts, using (3.2.3) and (3.2.31), we have

$$\int_0^t \int_0^1 v_x^2 dxds \leq C_1. \qquad (3.2.32)$$

By (3.1.3), integrating by parts and using Young's inequality gives

$$\int_0^1 v_x^2(x,t)dx + \int_0^t \int_0^1 (v_{xx}^2 + \theta^2 v_x^2 + u_x^2 v_x^2 + v_x^4)(x,s)dxds \leq C_1. \qquad (3.2.33)$$

Similarly, from (3.1.4), we infer that

$$\int_0^1 \theta_x^2(x,t)dx + \int_0^t \int_0^1 (\theta_{xx}^2 + \theta_x^2 u_x^2)(x,s)dxds \leq C_1. \qquad (3.2.34)$$

By (3.2.31)–(3.2.34) and (3.2.3), we obtain (3.2.4) and (3.2.6).

By (3.1.2)–(3.1.4) and (3.2.3), we observe that

$$\left| \frac{d}{dt} \int_0^1 \left(\mu \frac{u_x}{u} - v \right)^2 dx \right| \leq C_1 \int_0^1 \left(\theta_x^2 + u_x^2 + v^2 + \theta^2 u_x^2 \right) dx,$$

$$\left| \frac{d}{dt} \int_0^1 v_x^2 dx \right| \leq C_1 \int_0^1 \left(\theta_x^2 + \theta^2 u_x^2 + u_x^2 v_x^2 \right) dx,$$

$$\left| \frac{d}{dt} \int_0^1 \theta_x^2 dx \right| \leq C_1 \int_0^1 \left(\theta^2 v_x^2 + v_x^4 + \theta_x^2 u_x^2 \right) dx$$

which, along with (3.2.31)–(3.2.34) and (3.2.4), gives

$$\lim_{t \to +\infty} \int_0^1 \left(v_x^2 + \theta_x^2 + \left(\mu \frac{u_x}{u} - v \right)^2 \right) dx = 0. \tag{3.2.35}$$

In particular, since the boundary condition gives

$$|v(x,t)| \le \left(\int_0^1 v_x^2 dx \right)^{1/2}, \tag{3.2.36}$$

which, together with (3.2.35), yields

$$\lim_{t \to +\infty} \max_{x \in [0,1]} |v(x,t)| = 0. \tag{3.2.37}$$

Thus, in view of (3.2.35), we conclude

$$\lim_{t \to +\infty} \int_0^1 \left(v_x^2 + \theta_x^2 + u_x^2 \right)(x,t) dx = 0. \tag{3.2.38}$$

Now, for the Neumann boundary conditions, we have

$$|\theta(x,t) - \bar\theta| \le \left| \theta(x,t) - \int_0^1 \theta dx \right| + \frac{1}{2} \int_0^1 v^2(x,t) dx.$$

Employing (3.2.36) and (3.2.38), we see that in both cases

$$|\theta(x,t) - \bar\theta|^2 \le C_1 \int_0^1 (\theta_x^2 + v_x^2) dx, \tag{3.2.39}$$

and thus (3.2.35) implies

$$\lim_{t \to +\infty} \max_{x \in [0,1]} |\theta(x,t) - \bar\theta| = 0. \tag{3.2.40}$$

Finally, (3.1.7) yields

$$|u(x,t) - 1| \le \left| u(x,t) - \int_0^1 u dx \right| \le \left(\int_0^1 u_x^2 dx \right)^{1/2}, \tag{3.2.41}$$

and so recalling (3.2.37)–(3.2.38) and (3.2.40), we deduce that

$$\lim_{t \to +\infty} \max_{x \in [0,1]} \left(|v(x,t)| + |\theta(x,t) - \bar\theta| + |u(x,t) - 1| \right) = 0. \tag{3.2.42}$$

Set

$$V(t) := \int_0^1 \left(v_x^2 + \theta_x^2 \right) dx, \qquad D(x,t) := \int_0^1 \left(\mu \frac{u_x}{u} - v \right)^2 dx,$$

$$A(t) := \int_0^1 \left(\theta + \frac{1}{2} v^2 - \bar\theta \eta + \gamma \right) dx$$

where $\gamma = \bar{\theta}(\ln \theta - 1)$. Integrating the availability identity (3.2.11) over $[0, 1]$, we have

$$\frac{d}{dt}A(t) + \lambda V(t) \leq 0. \tag{3.2.43}$$

Observing the boundedness of θ, due to (3.2.40), it follows from the Taylor expansion of the function $\ln$, that

$$\lambda(\theta - \bar{\theta})^2 \leq (\theta - \bar{\theta}\ln\theta) + \gamma \leq \Lambda(\theta - \bar{\theta})^2. \tag{3.2.44}$$

Analogously, using (3.1.7), the boundedness of u, and the concavity of h,

$$\lambda \int_0^1 (u-1)^2 dx \leq -\int_0^1 h(u) dx \leq \Lambda \int_0^1 (u-1)^2 dx. \tag{3.2.45}$$

Adding (3.2.44) and (3.2.45) yields,

$$\lambda \int_0^1 \left((\theta - \bar{\theta})^2 + (u-1)^2 + v^2 \right) dx \leq A \leq \Lambda \int_0^1 \left((\theta - \bar{\theta})^2 + (u-1)^2 + v^2 \right) dx.$$

Hence, by (3.2.39), (3.2.41), and (3.2.36), we obtain

$$A(t) \leq \Lambda \int_0^1 (\theta_x^2 + v^2 + u_x^2) dx$$
$$\leq \Lambda \left[D(t) + \int_0^1 (\theta_x^2 + v^2) dx \right] \leq \Lambda(D(t) + V(t)). \tag{3.2.46}$$

In addition, from (3.2.36), (3.2.40) and Young's inequality, we see that

$$\frac{d}{dt}D(t) + \lambda D(t) \leq \Lambda V(t). \tag{3.2.47}$$

From (3.1.3)–(3.1.4), we infer

$$\frac{d}{dt}V(t) + \lambda \int_0^1 (v_{xx}^2 + \theta_{xx}^2) dx \leq \Lambda \left[V(t) + \int_0^1 (v_x^4 + u_x^2 + \theta_x^2 u_x^2 + u_x^2 v_x^2) dx \right]. \tag{3.2.48}$$

Noting the boundedness of $\int_0^1 (u_x^2 + v_x^2) dx$, the interpolation inequalities imply the integral on the right-hand side of (3.2.48) is estimated by

$$\lambda \int_0^1 (v_{xx}^2 + \theta_{xx}^2) dx + \Lambda \left(V(t) + \int_0^1 u_x^2 dx \right).$$

Thus, by (3.2.36),

$$\frac{d}{dt}V(t) \leq \Lambda(D(t) + V(t)). \tag{3.2.49}$$

Finally, multiplying (3.2.47) by a small constant $\varepsilon > 0$ and then adding the result to (3.2.43), we deduce

$$\frac{d}{dt}(A + \varepsilon D) + \lambda(D + V) \leq 0.$$

For sufficiently small ε, by (3.2.46) and (3.2.49) and the above inequality, we may conclude

$$\frac{d}{dt}(A + \varepsilon D + \varepsilon V) + \lambda(A + \varepsilon D + \varepsilon V) \leq 0.$$

Thus,

$$(A + \varepsilon D + \varepsilon V)(t) \leq \Lambda e^{-\lambda t}.$$

Recalling (3.2.36), (3.2.40) and (3.2.41), we deduce (3.2.7)–(3.2.8). And (3.2.5) follows from (3.2.7)–(3.2.8). The proof is complete.   □

Lemma 3.2.3. *Under the assumptions of Theorem 3.1.1, the following estimate holds:*

$$0 < C_1^{-1} \leq \theta(x,t), \quad \forall (x,t) \in [0,1] \times [0,+\infty). \tag{3.2.50}$$

Proof. To prove (3.2.50), we recall (3.2.8). Then we conclude that there exists a positive constant T, such that for any $t > T$,

$$\theta(x,t) \geq C_1^{-1} > 0. \tag{3.2.51}$$

For the lower bound of $\theta(x,t)$ on the interval $[0,T]$, we introduce $\omega = \frac{1}{\theta}$ and find that ω satisfies

$$\omega_t = \frac{\omega}{u^p}v_x - \mu\frac{\omega^2 v_x^2}{u} - 2\kappa\frac{\omega_x^2}{u\omega} + \kappa\left(\frac{\omega_x}{u}\right)_x. \tag{3.2.52}$$

Multiplying (3.2.52) by ω^{2p-1} and integrating the resultant over $(0,1)$, we obtain

$$\frac{1}{2p}\frac{d}{dt}\int_0^1 \omega^{2p}\,dx + \mu\int_0^1 \frac{\omega^{2p+1}v_x^2}{u}\,dx \leq \int_0^1 \frac{\omega^{2p}}{u^p}v_x\,dx$$

$$\leq \frac{\mu}{2}\int_0^1 \frac{\omega^{2p+1}v_x^2}{u}\,dx + C_1\int_0^1 \omega^{2p-1}\,dx. \tag{3.2.53}$$

Let $y(t) = \|\omega(\cdot,t)\|_{L^{2p}}$. Then we derive from (3.2.53) that

$$\frac{d}{dt}y(t) \leq C_1. \tag{3.2.54}$$

Hence,

$$\|\omega(\cdot,t)\|_{L^{2p}} \leq C_1(1+t). \tag{3.2.55}$$

Passing to limit, as $p \to +\infty$, we have

$$\|\omega(\cdot,t)\|_{L^\infty} \leq C_1(1+t), \tag{3.2.56}$$

i.e,

$$\theta(x,t) \geq C_1(1+t)^{-1} \geq C_1(1+T)^{-1}, \quad \forall t \in [0,T]$$

which, along with (3.2.51), gives (3.2.50).   □

In what follows we shall use the idea of [55] to prove exponential stability in H^1. Set

$$\Psi(u,\theta) = e(u,\theta) - \theta\eta(\theta,u) \qquad (3.2.57)$$

where $\eta(u,\theta)$ verifies

$$1 = c_v e_\theta = \theta\eta_\theta, \quad \eta_u = \mathcal{P}_\theta. \qquad (3.2.58)$$

Now we introduce the density of Newtonian fluid, $\rho = 1/u$, then $\eta = \eta(1/\rho,\theta)$ satisfies

$$\frac{\partial\eta}{\partial\rho} = \begin{cases} \frac{-\mathcal{P}_\theta}{\rho^2}, & p = 1 \\ \frac{1}{\mathcal{P}_\theta}, & p > 1 \end{cases}, \quad \frac{\partial\eta}{\partial\theta} = \frac{e_\theta}{\theta} = \frac{1}{\theta}. \qquad (3.2.59)$$

We consider the transform

$$A : (\rho,\theta) \in D_{\rho,\theta} = \{(\rho,\theta) : \rho > 0, \theta > 0\} \to (u,\eta) \in AD_{\rho,\theta}. \qquad (3.2.60)$$

Owing to the Jacobian $|\partial(u,\eta)/\partial(\rho,\theta)| = -e_\theta/\rho^2\theta < 0$ on $AD_{\rho,\theta}$, there is a unique inverse function $\theta = \theta(u,\eta)$ as the smooth function of $(u,\eta) \in AD_{\rho,\theta}$. Thus the functions $e, \mathcal{P}$ can also be regarded as the smooth functions of (u,η). We denote them by

$$e = e(u,\eta) :\equiv e(u,\theta(u,\eta)) = e(1/\rho,\theta),$$
$$\mathcal{P} = \mathcal{P}(u,\eta) :\equiv \mathcal{P}(u,\theta(u,\eta)) = \mathcal{P}(1/\rho,\theta). \qquad (3.2.61)$$

Then it follows from (3.2.57)–(3.2.61) that

$$e_u = -\mathcal{P}, \quad e_\eta = \theta,$$
$$\mathcal{P}_u = -(\rho^2\mathcal{P}_\rho + \theta\mathcal{P}_\theta^2/e_\theta), \quad \mathcal{P}_\eta = \theta\mathcal{P}_\theta/e_\theta,$$
$$\theta_u = -\theta\mathcal{P}_\theta/e_\theta, \quad \theta_\eta = \theta/e_\theta. \qquad (3.2.62)$$

We define the following energy form,

$$V(u,v,\eta) = \frac{v^2}{2} + e(u,\eta) - e(\bar{u},\bar{\eta}) - \frac{\partial e}{\partial u}(\bar{u},\bar{\eta})(u-\bar{u}) - \frac{\partial e}{\partial\eta}(\bar{u},\bar{\eta})(\eta-\bar{\eta}). \quad (3.2.63)$$

Lemma 3.2.4. *The unique global solution* $(u(t), v(t), \theta(t)) \in (H^1[0,1])^3$ *to problem* (3.1.2)–(3.1.6) *satisfies the following estimates:*

$$\frac{v^2}{2} + C_1^{-1}[(u-1)^2 + (\eta-\bar{\eta})^2] \leq V(u,v,\eta)$$

$$\leq \frac{v^2}{2} + C_1[(u-\bar{u})^2 + (\eta-\bar{\eta})^2]. \qquad (3.2.64)$$

Proof. The proof is similar to that of Lemma 2.2.2 (see also Qin [55]). ☐

Lemma 3.2.5. *There exist constants $C_1 > 0$ and $\gamma_1 = \gamma_1(C_1) > 0$ such that for any fixed $\gamma \in (0, \gamma_1]$, the global solution $(u(t), v(t), \theta(t)) \in H^1$ to problem (3.1.2)–(3.1.6) satisfies the following estimate:*

$$e^{\gamma t}\left(\|v(t)\|_{H^1}^2 + \|u(t) - 1\|_{H^1}^2 + \|\theta(t) - \bar{\theta}\|_{H^1}^2\right) \tag{3.2.65}$$

$$+ \int_0^t e^{\gamma t}\left(\|u_x\|^2 + \|v_x\|_{H^1}^2 + \|\theta_x\|_{H^1}^2 + \|v_t\|^2 + \|\theta_t\|^2\right)(\tau)d\tau \leq C_1, \quad \forall \, t > 0.$$

Proof. The proof is similar to that of Lemmas 2.2.3–2.2.4 (see also Qin [55]). $\square$

Lemma 3.2.6. *Under the assumptions of Theorem 3.1.1, the following estimates hold:*

$$\|v_t(t)\|^2 + \|\theta_t(t)\|^2 + \int_0^t (\|v_{tx}\|^2 + \|\theta_{tx}\|^2)(\tau)\,d\tau \leq C_2, \quad \forall t > 0, \tag{3.2.66}$$

$$\|u_{xx}(t)\|^2 + \int_0^t \|u_{xx}\|^2(\tau)\,d\tau \leq C_2, \quad \forall t > 0, \tag{3.2.67}$$

$$\|v_{xx}(t)\|^2 + \|\theta_{xx}(t)\|^2 + \int_0^t (\|v_{xxx}\|^2 + \|\theta_{xxx}\|^2)(\tau)\,d\tau \leq C_2, \quad \forall t > 0. \tag{3.2.68}$$

Proof. The proofs of (3.2.66) and (3.2.68) are similar to that of (2.3.1). We only need to prove (3.2.67). Differentiating (3.1.3) with respect to x, using (3.1.2) $(u_{txx} = v_{xxx})$, we see that

$$\mu\frac{\partial}{\partial t}\left(\frac{u_{xx}}{u}\right) - \mathcal{P}_u u_{xx} = v_{tx} + E(x, t) \tag{3.2.69}$$

where

$$E(x, t) = \mathcal{P}_{uu}u_x^2 + 2\mathcal{P}_{\theta u}\theta_x u_x + \mathcal{P}_{\theta\theta}\theta_x^2 + \mathcal{P}_\theta\theta_{xx} - 2\mu v_x u_x^2/u^3 + 2\mu u_x v_{xx}/u^2$$

$$= p(p+1)\frac{\theta}{u^{p+2}}u_x^2 - 2p\frac{\theta_x u_x}{u^{p+1}} + \frac{1}{u^{p+1}}\theta_{xx} - 2\mu\left(\frac{v_{xx}u_x}{u^2} - \frac{v_x u_x^2}{u^3}\right).$$

Multiplying (3.2.69) by u_{xx}/u, and by Young's inequality, by Lemmas 3.2.2–3.2.3, we can deduce that

$$\frac{d}{dt}\left\|\frac{u_{xx}}{u}\right\|^2 + C_1^{-1}\left\|\frac{u_{xx}}{u}\right\|^2$$

$$\leq \frac{1}{4C_1}\left\|\frac{u_{xx}}{u}\right\|^2 + C_1\left(\|\theta_x\|_{L^\infty}^2\|u_x\|^2 + \|u_x\|_{L^4}^4 + \|v_{tx}\|^2\right.$$

$$\left. + \|u_x\|_{L^\infty}^2\|v_{xx}\|^2 + \|\theta_{xx}\|^2 + \|v_x\|_{L^\infty}^2\|u_x\|_{L^4}^4\right)$$

$$\leq \frac{1}{2C_1}\left\|\frac{u_{xx}}{u}\right\|^2 + C_2\left(\|\theta_x\|^2 + \|\theta_{xx}\|^2 + \|u_x\|^2 + \|v_{tx}\|^2\right) \tag{3.2.70}$$

which, combined with Lemma 3.2.2, gives (3.2.67). $\square$

Lemma 3.2.7. *There exist constants $C_2 > 0$ and $\gamma_2 = \gamma_2(C_2) > 0$ such that for any fixed $\gamma \in (0, \gamma_2]$, the global solution $(u(t), v(t), \theta(t)) \in H_+^2$ to problem (3.1.2)–(3.1.6) satisfies that the following estimates:*

$$e^{\gamma t} \left(\|u(t) - 1\|_{H^2}^2 + \|v(t)\|_{H^2}^2 + \|\theta(t) - \bar{\theta}\|_{H^2}^2 \right)$$

$$+ \int_0^t e^{\gamma \tau} \left(\|u - 1\|_{H^2}^2 + \|v\|_{H^3}^2 + \|\theta - \bar{\theta}\|_{H^3}^2 \right) (\tau) d\tau \leq C_2, \quad \forall t > 0, \quad (3.2.71)$$

$$e^{\gamma t} \left(\|v_t(t)\|^2 + \|\theta_t(t)\|^2 \right) + \int_0^t e^{\gamma \tau} \left(\|v_{tx}\|^2 + \|\theta_{tx}\|^2 \right) (\tau) d\tau \leq C_2, \quad \forall t > 0.$$

$$(3.2.72)$$

Proof. The proof is similar to that of Lemma 2.3.4. $\qquad \square$

Proof of Theorem 3.1.1. By Lemmas 3.2.1–3.2.7, we complete the proof of Theorem 3.1.1. $\qquad \square$

3.3 Proof of Theorem 3.1.2

In this section, we shall complete the proof of Theorem 3.1.2 and take the assumptions in Theorem 3.1.2 to be valid. We begin with the following lemma.

Lemma 3.3.1. *Under assumptions of Theorem 3.1.2, for any $(u_0, v_0, \theta_0) \in H_+^4$, we have*

$$\|v_{tx}(x, 0)\| + \|\theta_{tx}(x, 0)\| \leq C_3, \quad (3.3.1)$$

$$\|v_{tt}(x, 0)\| + \|\theta_{tt}(x, 0)\| + \|v_{txx}(x, 0)\| + \|\theta_{txx}(x, 0)\| \leq C_4, \quad (3.3.2)$$

$$\|v_{tt}(t)\|^2 + \int_0^t \|v_{ttx}(\tau)\|^2 \, d\tau \leq C_4 + C_4 \int_0^t \|\theta_{txx}(\tau)\|^2 \, d\tau, \quad \forall t > 0, \quad (3.3.3)$$

$$\|\theta_{tt}(t)\|^2 + \int_0^t \|\theta_{ttx}(\tau)\|^2 \, d\tau \quad (3.3.4)$$

$$\leq C_4 + C_2 \varepsilon^{-1} \int_0^t \|\theta_{txx}(\tau)\|^2 \, d\tau + C_1 \varepsilon \int_0^t \left(\|v_{ttx}\|^2 + \|v_{txx}\|^2 \right) (\tau) \, d\tau, \quad \forall t > 0$$

for $\varepsilon > 0$ small enough.

Proof. The proof is similar to that of Lemma 2.4.1. $\qquad \square$

Lemma 3.3.2. *Under assumptions of Theorem 3.1.2, for any $(u_0, v_0, \theta_0) \in H_+^4$, the following estimates hold for any $t > 0$ and for $\varepsilon \in (0, 1)$ small enough,*

$$\|v_{tx}(t)\|^2 + \int_0^t \|v_{txx}(\tau)\|^2 \, d\tau \leq C_4 + C_1 \varepsilon^2 \int_0^t (\|v_{ttx}\|^2 + \|\theta_{txx}\|^2)(\tau) \, d\tau, \quad (3.3.5)$$

$$\|\theta_{tx}(t)\|^2 + \int_0^t \|\theta_{txx}(\tau)\|^2 \, d\tau \leq C_4 + C_2 \varepsilon^2 \int_0^t \|v_{txx}(\tau)\|^2 \, d\tau. \quad (3.3.6)$$

Proof. The proof is similar to that of Lemma 2.4.2. $\qquad \square$

Lemma 3.3.3. *Under assumptions of Theorem 3.1.2, for any* $(u_0, v_0, \theta_0) \in H_+^4$, *we have for any* $t > 0$,

$$\|v_{tt}(t)\|^2 + \|v_{tx}(t)\|^2 + \|\theta_{tt}(t)\|^2 + \|\theta_{tx}(t)\|^2$$

$$+ \int_0^t \left(\|v_{ttx}\|^2 + \|v_{txx}\|^2 + \|\theta_{ttx}\|^2 + \|\theta_{txx}\|^2 \right)(\tau)\, d\tau \le C_4, \qquad (3.3.7)$$

$$\|u_{xxx}(t)\|_{H^1}^2 + \|v_{xxx}(t)\|_{H^1}^2 + \|\theta_{xxx}(t)\|_{H^1}^2 + \|v_{txx}(t)\|^2 + \|\theta_{txx}(t)\|^2$$

$$+ \int_0^t \left(\|v_{tt}\|^2 + \|\theta_{tt}\|^2 + \|v_{txx}\|_{H^1}^2 + \|\theta_{txx}\|_{H^1}^2 \right)(\tau)\, d\tau \le C_4, \qquad (3.3.8)$$

$$\int_0^t (\|u_{xxx}\|_{H^1}^2 + \|v_{xxxx}\|_{H^1}^2 + \|\theta_{xxxx}\|_{H^1}^2)(\tau)\, d\tau \le C_4. \qquad (3.3.9)$$

Proof. The proof is similar to that of Lemma 2.4.3.   □

Lemma 3.3.4. *Under assumptions of Theorem 3.1.2, for any* $(u_0, v_0, \theta_0) \in H_+^4$, *there exists a positive constant* $\gamma_4^{(1)} = \gamma_4^{(1)}(C_4) \le \gamma_2(C_2)$ *such that for any fixed* $\gamma \in (0, \gamma_4^{(1)}]$, *the following estimates hold for any* $t > 0$ *and* $\varepsilon \in (0, 1)$ *small enough,*

$$e^{\gamma t}\|v_{tt}(t)\|^2 + \int_0^t e^{\gamma \tau}\|v_{ttx}(\tau)\|^2\, d\tau \le C_4 + C_2 \int_0^t e^{\gamma \tau}(\|\theta_{txx}\|^2 + \|v_{txx}\|^2)(\tau)\, d\tau,$$

$$(3.3.10)$$

$$e^{\gamma t}\|\theta_{tt}(t)\|^2 + \int_0^t e^{\gamma \tau}\|\theta_{ttx}(\tau)\|^2\, d\tau \le C_4 \varepsilon^{-3} + C_2 \varepsilon^{-1} \int_0^t e^{\gamma \tau}\|\theta_{txx}(\tau)\|^2\, d\tau$$

$$+ \varepsilon \int_0^t e^{\gamma \tau}(\|v_{txx}\|^2 + \|v_{ttx}\|^2)(\tau)\, d\tau.$$

$$(3.3.11)$$

Proof. The proof is similar to that of Lemma 2.4.5.   □

Lemma 3.3.5. *Under assumptions of Theorem 3.1.2, for any* $(u_0, v_0, \theta_0) \in H_+^4$, *there exists a positive constant* $\gamma_4^{(2)} \le \gamma_4^{(1)}$ *such that for any fixed* $\gamma \in (0, \gamma_4^{(2)}]$, *the following estimates hold for any* $t > 0$ *and* $\varepsilon \in (0, 1)$ *small enough,*

$$e^{\gamma t}\|v_{tx}(t)\|^2 + \int_0^t e^{\gamma \tau}\|v_{txx}(\tau)\|^2\, d\tau \le C_4 + C_2 \varepsilon^2 \int_0^t e^{\gamma \tau}(\|v_{ttx}\|^2 + \|\theta_{txx}\|^2)(\tau)\, d\tau,$$

$$(3.3.12)$$

$$e^{\gamma t}\|\theta_{tx}(t)\|^2 + \int_0^t e^{\gamma \tau}\|\theta_{txx}(\tau)\|^2\, d\tau \le C_4 + \varepsilon^2 \int_0^t e^{\gamma \tau}\|v_{txx}(\tau)\|^2\, d\tau, \qquad (3.3.13)$$

$$e^{\gamma t}(\|v_{tx}(t)\|^2 + \|\theta_{tx}(t)\|^2) + \int_0^t e^{\gamma \tau}(\|v_{txx}\|^2 + \|\theta_{txx}\|^2)(\tau)\, d\tau$$

$$\le C_4 + C_2 \varepsilon^2 \int_0^t e^{\gamma \tau}\|v_{ttx}(\tau)\|^2\, d\tau. \qquad (3.3.14)$$

Proof. The proof is similar to that of Lemma 2.4.6.   □

Lemma 3.3.6. *Under assumptions of Theorem* 3.1.2, *for any* $(u_0, v_0, \theta_0) \in H_+^4$, *there is a positive constant* $\gamma_4 \leq \gamma_4^{(2)}$ *such that for any fixed* $\gamma \in (0, \gamma_4]$, *the following estimates hold for any* $t > 0$,

$$
e^{\gamma t}(\|v_{tt}(t)\|^2 + \|\theta_{tt}(t)\|^2 + \|v_{tx}(t)\|^2 + \|\theta_{tx}(t)\|^2)
$$

$$
+ \int_0^t e^{\gamma \tau}\Big(\|v_{ttx}\|^2 + \|\theta_{ttx}\|^2 + \|v_{txx}\|^2 + \|\theta_{txx}\|^2\Big)(\tau)\, d\tau \leq C_4, \qquad (3.3.15)
$$

$$
e^{\gamma t}\|u_{xxx}(t)\|_{H^1}^2 + \int_0^t e^{\gamma \tau}\|u_{xxx}(\tau)\|_{H^1}^2\, d\tau \leq C_4, \qquad (3.3.16)
$$

$$
e^{\gamma t}\Big(\|v_{xxx}(t)\|_{H^1}^2 + \|\theta_{xxx}(t)\|_{H^1}^2 + \|v_{txx}(t)\|^2 + \|\theta_{txx}(t)\|^2\Big)
$$

$$
+ \int_0^t e^{\gamma \tau}\Big(\|v_{xxxx}\|_{H^1}^2 + \|\theta_{xxxx}\|_{H^1}^2 + \|v_{txx}\|_{H^1}^2
$$

$$
+ \|\theta_{txx}\|_{H^1}^2 + \|v_{tt}\|^2 + \|\theta_{tt}\|^2\Big)(\tau)\, d\tau \leq C_4. \qquad (3.3.17)
$$

Proof. The proof is similar to that of Lemmas 2.4.7–2.4.8.   □

Proof of Theorem 3.1.2. By Lemmas 3.3.1–3.3.6, Theorem 3.1.1 and Sobolev's embedding theorem, we complete the proof of Theorem 3.1.2.   □

3.4 Bibliographic Comments

Now let's first recall some previous works in this direction. For the case of an ideal gas, i.e.,

$$
e = C_v \theta, \quad \mathcal{P} = R\frac{\theta}{u} \qquad (3.4.1)
$$

with suitable positive constants C_v and R, the global existence and asymptotic behavior of smooth (generalized) solutions in $H^i(i = 1, 2)$ to the system (3.1.2)–(3.1.4) have been investigated by many authors (see, e.g., Antontsev, Kazhikhov and Monakhov [1], Chen, Hoff and Trivisa [4], Hoff [22], Hsiao and Luo [24], Jiang [26, 27], Kawashima and Nishida [31], Kawohl [32], Matsumura and Nishida [40, 41, 42, 43], Nagasawa [45], Okada and Kawashima [47], Qin [59, 50, 49, 51, 52], Qin and Muñoz Rivera [66], Qin, Wu and Liu [73]) on the initial boundary value problems. For the Cauchy problem with (3.4.1), we refer to the works Itaya [25], Kawashima and Nishida [31], Kazhikhov [35], Kazhikhov and Shelukhin [36], Matsumura and Nishida [40, 41], Okada and Kawashima [47], Qin [55], Zheng and Shen [85].

For a nonlinear one-dimensional heat-conductive viscous real gas with some constitutive equations and special forms of functions μ, κ and $\mathcal{P}$, the classical solutions (weak solutions in H^1) exist globally in time and converge exponentially to a steady state in H^1 with some strong assumptions for (3.1.2)–(3.1.6) (see, Jiang

[26, 27], Kawohl [32]). Later on, Qin [49, 50, 51, 52, 54] established the same results as above on the global existence and exponential stability with some weaker assumptions. For multidimensional initial boundary value problems and Cauchy problems, the global existence and asymptotic behavior of smooth solutions have been investigated for general domains only in case of "small initial data" (see, e.g, Antontsev, Kazhikhov and Monakhov [1], Deckelnick [9], Fujita-Yashima and Benabidallah [80, 81], Hoff [23], Itaya [25], Kawashima and Nishida [31], Kazhikhov [35], Kazhikhov and Shelukhin [36], Matsumura and Nishida [40, 41, 42, 43], Okada and Kawashima [47], Qin [59, 55], Zheng and Shen [85] and references cited therein). However, in our case, the form of the pressure $\mathcal{P}$ in (3.1.1) can be regarded as a generalization of the constitutive equation for an ideal gas, as well as a modification of the relation for a barotropic gas, where $\mathcal{P} = u^{-p}$. From this point of view, (3.1.1) is an interpolation between these models.

In this direction, based on the result obtained in [37] with $\mathcal{P} = \frac{\theta}{u^p}$, we have established in this chapter the regularity and exponential stability of global solutions in H^i $(i = 2, 4)$, which are two new ingredients of this chapter. As a result, by the embedding theorem, the global solution obtained in H^4 is actually a classical one in $C^{3+1/2}$ when it is subjected to the corresponding compatibility conditions. Thus the exponential stability of classical solutions is obtained, which is a new result for this model. The aim of this chapter is to prove the global existence and exponential of solutions in H^i $(i = 2, 4)$ for equations (3.1.2)–(3.1.4) for boundary conditions (3.1.6) with pressure (3.1.1).

Chapter 4

Global Existence and Exponential Stability for the pth Power Viscous Reactive Gas

4.1 Introduction

In this chapter, we prove the global existence and exponential stability of solutions in $H^i (i = 2, 4)$ for the compressible Navier-Stokes equations, which arise in the study of a thermal explosion and describe the dynamic combustion for a reactive Newtonian fluid, confined between two infinite parallel plates. We assume that the pressure $\mathcal{P}$, in terms of the absolute temperature θ and the specific volume u, is given by

$$\mathcal{P} = \frac{\theta}{u^p} \tag{4.1.1}$$

with the pressure exponent $p \geq 1$.

The balance laws of mass, momentum, and energy, coupled with the description of the chemical reaction for one-dimensional (in Lagrangian form) case are

$$u_t = v_x, \tag{4.1.2}$$

$$v_t = \left(-\mathcal{P} + \mu \frac{v_x}{u} \right)_x =: \sigma_x, \tag{4.1.3}$$

$$c_v \theta_t = \left(-\mathcal{P} + \mu \frac{v_x}{u} \right) v_x + \left(\kappa \frac{\theta_x}{u} \right)_x + \delta f(u, \theta, z), \tag{4.1.4}$$

$$z_t = \left(d \frac{z_x}{u^2} \right)_x - f(u, \theta, z). \tag{4.1.5}$$

Here, u, v, θ, z, f are the specific volume, velocity, the absolute temperature, the mass fraction of the unburned fuel, the function of chemical reaction, respectively. All the above quantities are assumed to vary spatially only in the direction perpendicular to the plates. The positive constants c_v, μ, κ, d, δ represent specific heat, viscosity, conductivity, the species diffusion coefficient, and the reactive rate, respectively. Since the magnitude of the specific heat c_v plays no role in the mathematical analysis of the system, in what follows we will assume the scaling

$$c_v = 1.$$

The function f describes the intensity of the chemical reaction, typically one has

$$f(u, \theta, z) = \varepsilon u^{1-m} z^m \exp \frac{\theta - 1}{\varepsilon \theta} \qquad (4.1.6)$$

which is called the Arrhenius rate law for chemical reaction. Here, ε^{-1}, m are positive constants and denote the activation energy and the overall sum of the individual reaction orders for the fuel and oxidizer, respectively. The physically interesting case is that $m \geq 1$.

We consider a typical initial boundary value problem for (4.1.2)–(4.1.6) in the reference domain $\{(x, t) : 0 < x < 1, t \geq 0\}$ under the initial conditions and boundary conditions

$$u(x, 0) = u_0(x), \; v(x, 0) = v_0(x), \; \theta(x, 0) = \theta_0(x), \; z(x, 0) = z_0(x), \qquad (4.1.7)$$

$$v(0, t) = v(1, t) = 0, \; \theta_x(0, t) = \theta_x(1, t) = 0, \; z_x(0, t) = z_x(1, t) = 0. \qquad (4.1.8)$$

The physical meaning of the boundary conditions (4.1.8) are clear, namely, both ends of the rods are clamped, thermally insulated and impermeable.

The notation in this chapter is standard. We put $\|\cdot\| = \|\cdot\|_{L^2[0,1]}$. Subscripts t and x denote the (partial) derivatives with respect to t and x, respectively. We use C_i $(i = 1, 2, 4)$ to denote the generic positive constant depending on the $H^i[0, 1]$ norm of initial data $(u_0, v_0, \theta_0, z_0)$, $\min_{x \in [0,1]} u_0(x)$, $\min_{x \in [0,1]} \theta_0(x)$ and $\min_{x \in [0,1]} z_0(x)$, but independent of variable t. C stands for the absolute positive constant independent of C_i $(i = 1, 2, 4)$ and initial data.

For convenience and without loss of generality, we may assume $\int_0^1 u_0(x) \, dx = 1$. Then from conservation of mass and boundary condition (4.1.8), we have

$$\int_0^1 u(x, t) \, dx = \int_0^1 u_0(x) \, dx = 1. \qquad (4.1.9)$$

In Chapter 3, we discussed the case $\delta = 0$ (without chemical reaction). In this chapter, we shall investigate the case $\delta > 0$ and the contribution of a chemical reaction to the existence and exponential stability of global solutions. Hence we have improved the previous results in [37]. Due to the involvement of a chemical reaction, the present situation is more complicated than that in Chapter 3, and more delicate and careful analyses are needed.

Now we are in a position to state our main results which are chosen from [62].

Define

$$H_+^1 = \{(u, v, \theta, z) \in (H^1[0,1])^4 : u(x) > 0, \theta(x) > 0, z(x) \geq 0, \forall x \in [0,1],$$
$$v(0) = v(1) = 0\},$$
$$H_+^i = \{(u, v, \theta, z) \in (H^i[0,1])^4 : u(x) > 0, \theta(x) > 0, z(x) \geq 0, \forall x \in [0,1],$$
$$v(0) = v(1) = \theta'(0) = \theta'(1) = z'(0) = z'(1) = 0\}, \ i = 2, 4.$$

Theorem 4.1.1. *Suppose that $(u_0, v_0, \theta_0, z_0) \in H_+^2$ and the compatibility conditions hold. Then there exists a unique global solution $(u(t), v(t), \theta(t), z(t)) \in H_+^2$ to the problem (4.1.2)–(4.1.8) verifying that for any $(x, t) \in [0,1] \times [0, +\infty)$,*

$$0 < C_1^{-1} \leq u(x,t) \leq C_1, \quad 0 < C_1^{-1} \leq \theta(x,t) \leq C_1, \quad 0 \leq z(x,t) \leq \max_{x \in [0,1]} z_0(x)$$
$$(4.1.10)$$

and for any $t > 0$,

$$\|u(t) - 1\|_{H^2}^2 + \|v(t)\|_{H^2}^2 + \|\theta(t) - \bar{\theta}\|_{H^2}^2 + \|z(t)\|_{H^2}^2 + \|v_t(t)\|^2 + \|\theta_t(t)\|^2 + \|z_t(t)\|^2$$
$$+ \int_0^t (\|u - 1\|_{H^2}^2 + \|v\|_{H^3}^2 + \|\theta - \bar{\theta}\|_{H^3}^2 + \|z_x\|_{H^2}^2 + \|v_t\|_{H^1}^2$$
$$+ \|\theta_t\|_{H^1}^2 + \|z_t\|_{H^1}^2)(\tau) d\tau \leq C_2.$$
$$(4.1.11)$$

Moreover, there are constants $C_2 > 0$ and $\gamma_2 = \gamma_2(C_2) > 0$ such that for any fixed $\gamma \in (0, \gamma_2]$, the following estimate holds for any $t > 0$,

$$e^{\gamma t} \Big(\|u(t) - 1\|_{H^2}^2 + \|v(t)\|_{H^2}^2 + \|\theta(t) - \bar{\theta}\|_{H^2}^2$$
$$+ \|z(t)\|_{H^2}^2 + \|v_t(t)\|^2 + \|\theta_t(t)\|^2 + \|z_t(t)\|^2 \Big)$$
$$+ \int_0^t e^{\gamma \tau} \Big(\|u - 1\|_{H^2}^2 + \|v\|_{H^3}^2$$
$$+ \|\theta - \bar{\theta}\|_{H^3}^2 + \|z\|_{H^3}^2 + \|v_{tx}\|^2 + \|\theta_{tx}\|^2 + \|z_{tx}\|^2 \Big)(\tau) d\tau \leq C_2 \quad (4.1.12)$$

where $\bar{\theta} = \int_0^1 \left(\theta_0 + \frac{1}{2}v_0^2 + \delta z_0\right) dx$.

Theorem 4.1.2. *Suppose that $(u_0, v_0, \theta_0, z_0) \in H_+^4$ and the compatibility conditions hold. Then there exists a unique global solution $(u(t), v(t), \theta(t), z(t)) \in H_+^4$ to the problem (4.1.2)–(4.1.8) such that for any $t > 0$,*

$$\|u(t) - 1\|_{H^4}^2 + \|v(t)\|_{H^4}^2 + \|\theta(t) - \bar{\theta}(t)\|_{H^4}^2 + \|z(t)\|_{H^4}^2 + \|v_t(t)\|_{H^2}^2$$
$$+ \|v_{tt}(t)\|^2 + \|\theta_t(t)\|_{H^2}^2 + \|\theta_{tt}(t)\|^2 + \|z_t(t)\|_{H^2}^2 + \|z_{tt}(t)\|^2$$
$$+ \int_0^t \Big(\|u - 1\|_{H^4}^2 + \|v\|_{H^5}^2 + \|\theta - \bar{\theta}\|_{H^5}^2 + \|z\|_{H^5}^2 + \|v_t\|_{H^3}^2 + \|\theta_t\|_{H^3}^2$$
$$+ \|z_t\|_{H^3}^2 + \|v_{tt}\|_{H^1}^2 + \|\theta_{tt}\|_{H^1}^2 + \|z_{tt}\|_{H^1}^2 \Big)(\tau) d\tau \leq C_4.$$
$$(4.1.13)$$

Moreover, there are constants $C_4 > 0$ and $\gamma_4 = \gamma_4(C_4) > 0$ such that for any fixed $\gamma \in (0, \gamma_4]$, the following estimate holds for any $t > 0$,

$$e^{\gamma t}\Big(\|u(t) - 1\|_{H^4}^2 + \|v(t)\|_{H^4}^2 + \|\theta(t) - \bar{\theta}\|_{H^4}^2 + \|z(t)\|_{H^4}^2 + \|v_t(t)\|_{H^2}^2$$

$$+ \|v_{tt}(t)\|^2 + \|\theta_t(t)\|_{H^2}^2 + \|\theta_{tt}(t)\|^2 + \|z_t(t)\|_{H^2}^2 + \|z_{tt}(t)\|^2\Big)$$

$$+ \int_0^t e^{\gamma \tau}\Big(\|u - 1\|_{H^4}^2 + \|v\|_{H^5}^2 + \|\theta - \bar{\theta}\|_{H^5}^2 + \|z_x\|_{H^4}^2 + \|v_t\|_{H^3}^2 + \|\theta_t\|_{H^3}^2$$

$$+ \|z_t\|_{H^3}^2 + \|v_{tt}\|_{H^1}^2 + \|\theta_{tt}\|_{H^1}^2 + \|z_{tt}\|_{H^1}^2\Big)(\tau)\, d\tau \le C_4. \qquad (4.1.14)$$

Remark 4.1.1. Obviously, it is easy to see that similar results hold for the boundary conditions

$$v(0,t) = v(1,t) = 0, \ \theta(0,t) = \theta(1,t) = \text{const.} > 0, \ z_x(0,t) = z_x(1,t) = 0.$$

Remark 4.1.2. Theorems 4.1.1–4.1.2 have improved the results in [38].

4.2   Global Existence in H^2

In this section we establish the global existence in H^2 by a series of lemmas. The next lemma concerns the estimate in H_+^1.

Lemma 4.2.1. *Under the conditions in Theorem 4.1.1, the following estimates hold:*

$$0 < C_1^{-1} \le u(x,t) \le C_1,$$

$$0 < C_1^{-1} \le \theta(x,t) \le C_1, \ \forall (x,t) \in [0,1] \times [0, +\infty), \qquad (4.2.1)$$

$$0 \le z(x,t) \le \max_{x \in [0,1]} z_0(x), \quad (x,t) \in [0,1] \times [0, +\infty), \qquad (4.2.2)$$

$$\|u(t) - 1\|_{H^1}^2 + \|v(t)\|_{H^1}^2 + \|\theta(t) - \bar{\theta}\|_{H^1}^2 + \|z(t)\|_{H^1}^2 + \|f(t)\|^2 \qquad (4.2.3)$$

$$+ \int_0^t (\|u - 1\|_{H^1}^2 + \|v\|_{H^2}^2 + \|\theta - \bar{\theta}\|_{H^2}^2 + \|f\|_{L^1} + \|z_x\|_{H^1}^2)(\tau)\, d\tau \le C_1.$$

Moreover, as $t \to +\infty$, we have

$$\|u_x(t)\|^2 + \|v_x(t)\|^2 + \|\theta_x(t)\|^2 + \|z_x(t)\|^2 \to 0, \qquad (4.2.4)$$

$$\max_{x \in [0,1]} \big(|u(x,t) - 1| + |v(x,t)| + |\theta(x,t) - \bar{\theta}| + |z(x,t)|\big) \to 0. \qquad (4.2.5)$$

Proof. Set $z_-(x,t) = \min\{z(x,t), 0\}$ and $z_+(x,t) = \max\{z(x,t), \max_{y \in [0,1]} z_0(y)\}$. In order to prove (4.2.2), we multiply (4.1.5) by z_+ and integrate in space, obtaining

$$\frac{1}{2}\frac{d}{dt}\int_0^1 z_+^2\, dx = -d\int_0^1 \frac{(z_+)_x^2}{u^2}\, dx - \int_0^1 f z_+\, dx \le 0.$$

In view of the initial condition $\int_0^1 z_+^2(x,0) = 0$, we conclude that $\int_0^1 z_+^2 dx = 0$. So we obtain (4.2.2). We note that the entropy η of a pth power viscous reactive gas is a concave function

$$\eta(u,\theta) = \log \theta + h(u), \tag{4.2.6}$$

where

$$h(u) = \begin{cases} \log u, & p = 1, \\ \frac{1}{p-1}(1 - u^{1-p}), & p > 1, \end{cases}$$

which satisfies the following standard entropy identity:

$$\eta_t = \mu \frac{v_x^2}{u\theta} + \kappa \frac{\theta_x^2}{u\theta^2} + \delta \frac{f}{\theta} - \left(\frac{q}{\theta}\right)_x. \tag{4.2.7}$$

Thus,

$$\frac{d}{dt} \int_0^1 \left(\theta + \frac{1}{2}v^2 + \delta z - \bar{\theta}\eta\right) dx = -\bar{\theta} \int_0^1 \left(\mu \frac{v_x^2}{u\theta} + \kappa \frac{\theta_x^2}{u\theta^2} + \delta \frac{f}{\theta}\right) dx. \tag{4.2.8}$$

Integrating (4.2.8) over $[0,t]$, we get

$$\int_0^1 \left(\theta + \frac{1}{2}v^2 + \delta z - \bar{\theta}\eta\right) dx + \bar{\theta} \int_0^1 \left(\mu \frac{v_x^2}{u\theta} + \kappa \frac{\theta_x^2}{u\theta^2} + \delta \frac{f}{\theta}\right) dx \leq C_1. \tag{4.2.9}$$

Integrating (4.2.6) in space and then using Jensen's inequality, we receive

$$\int_0^1 \eta dx \leq \log\left(\int_0^1 \theta dx\right) + h\left(\int_0^1 u dx\right) = \log\left(\int_0^1 \theta dx\right).$$

Thus, in view of (4.2.9), we see that

$$\int_0^1 \left(\theta + \frac{1}{2}v^2 + \delta z\right) dx + \bar{\theta} \int_0^1 \left(\mu \frac{v_x^2}{u\theta} + \kappa \frac{\theta_x^2}{u\theta^2} + \delta \frac{f}{\theta}\right) dx \leq C_1 + \bar{\theta}\log\left(\int_0^1 \theta dx\right). \tag{4.2.10}$$

In particular,

$$\int_0^1 \theta dx \leq C_1 + \bar{\theta}\log\left(\int_0^1 \theta dx\right),$$

which yields

$$0 < C_1^{-1} \leq \int_0^1 \theta dx \leq C_1. \tag{4.2.11}$$

Using (4.2.11) in (4.2.10), we establish

$$\int_0^1 v^2(x,t)dx + \int_0^t \int_0^1 \left(\mu \frac{v_x^2}{u\theta} + \kappa \frac{\theta_x^2}{u\theta^2} + \delta \frac{f}{\theta}\right)(x,\tau)dxd\tau \leq C_1. \tag{4.2.12}$$

Since the proof of the pointwise bound on u does not involve reactions terms, it can be carried out exactly as in the proof of (3.2.5) in Chapter 3, and thus we omit it.

Multiply (4.1.5) by z and integrate in space and in time to get

$$\int_0^1 z^2(x,t)dx + \int_0^t \int_0^1 \frac{z_x^2}{u^2}(x,\tau)dxd\tau \le C_1. \tag{4.2.13}$$

The following estimate was proved in Lemma 3.2.1 in Chapter 3,

$$\int_0^1 (u_x^2 + v_x^2)dx + \int_0^t \int_0^1 (u_x^2 + v_x^2 + \theta_x^2 + v_{xx}^2 + \theta^2 u_x^2 + u_x^2 v_x^2 + v_x^4)(x,s)dxds \le C_1. \tag{4.2.14}$$

Our next goal is to prove the bounds involving the reaction equation (4.1.5) and the conservation of energy (4.1.4). Note the following simple consequence of (4.1.4), (4.1.8) and (4.2.12):

$$\int_0^t \int_0^1 f dxds = -\int_0^t \int_0^1 z_t dxds = \int_0^1 z_0 dx - \int_0^1 z(x,t)dx \le C_1.$$

Thus,

$$\int_0^t \int_0^1 f dxds \le C_1, \quad 0 \le f(u,\theta,z)(x,t) \le C_1. \tag{4.2.15}$$

Next, from (4.1.4), integration by parts and Young's inequality, we get the boundary conditions

$$\frac{d}{dt}\left(\int_0^1 \theta_x^2\right) = 2\int_0^1 \theta_x \theta_{xt} dx = -2\int_0^1 \theta_t \theta_{xx} dx$$

$$\le C_1 \int_0^1 (|\theta v_x \theta_{xx}| + |v_x^2 \theta_{xx}| + |\theta_x u_x \theta_{xx}| + |f\theta_{xx}|)dx - C_1^{-1}\int_0^1 \theta_{xx}^2 dx$$

$$\le C_1 \int_0^1 (\theta^2 v_x^2 + v_x^4 + \theta_x^2 u_x^2 + f^2)dx - C_1^{-1}\int_0^1 \theta_{xx}^2 dx. \tag{4.2.16}$$

Note the following interpolation inequality:

$$\max_{x\in[0,1]} \theta_x^2(x,t) \le C_1 \int_0^1 \theta_x^2(x,t)dx + C_1 \int_0^1 \theta_{xx}^2(x,t)dx. \tag{4.2.17}$$

Thus, in view of (4.2.14), we have

$$\int_0^t \int_0^1 \theta_x^2 u_x^2 dxds \le \int_0^t \max_{x\in[0,1]} \theta_x^2(x,s)\left(\int_0^1 u_x^2 dx\right)ds$$

$$\le C_1 + C_1 \int_0^t \int_0^1 \theta_{xx}^2 dxds. \tag{4.2.18}$$

Now, integrating (4.2.16) over $[0,t]$ and noting (4.2.15), (4.2.18), we receive

$$\int_0^1 \theta_x^2 dx + \int_0^t \int_0^1 (\theta_{xx}^2 + \theta_x^2 u_x^2)dxds \le C_1. \tag{4.2.19}$$

In a similar manner, by (4.1.5), the boundary condition (4.1.7) and Young's inequality, we obtain

$$\frac{d}{dt}\left(\int_0^1 z_x^2 dx\right) = 2\int_0^1 z_x z_{xt} dx = -2\int_0^1 z_t z_{xx} dx$$

$$= -2d\int_0^1 z_{xx}\left(\frac{z_x}{u^2}\right)_x dx + 2\int_0^1 f z_{xx} dx$$

$$\leq C_1\int_0^1 (z_x^2 u_x^2 + f^2)dx - C_1^{-1}\int_0^1 z_{xx}^2 dx. \qquad (4.2.20)$$

Again, since

$$\max_{x\in[0,1]} z_x^2(x,t) \leq C_1\int_0^1 z_x^2(x,t)dx + C_1\int_0^1 z_{xx}^2 dx,$$

in view of (4.2.13) it follows that

$$\int_0^t\int_0^1 z_x^2 u_x^2 dx ds \leq \int_0^t \max_{x\in[0,1]} z_x^2(x,s)\left(\int_0^1 u_x^2 dx\right) ds \leq C_1 + C_1\int_0^t\int_0^1 z_{xx}^2 dx ds. \qquad (4.2.21)$$

Upon integrating (4.2.20) in time and inserting (4.2.21), (4.2.15)–(4.2.19), we have

$$\int_0^1 z_x^2 dx + \int_0^t\int_0^1 (z_{xx}^2 + z_x^2 u_x^2)dx ds \leq C_1. \qquad (4.2.22)$$

By (4.2.14), (4.2.15), (4.2.19), (4.2.22), we obtain (4.2.3).

We will prove (4.2.4). It is sufficient to show that the following functions and their derivatives are integrable in time: $\int_0^1 u_x^2(x,t)dx$, $\int_0^1 v_x^2(x,t)dx$, $\int_0^1 z_x^2 dx$. The integrability of the mentioned functions is stated in (4.2.14) and (4.2.13). The integrability of the derivatives of $\int_0^1 \theta_x^2(x,t)dx$ and $\int_0^1 z_x^2(x,t)dx$ is a consequence of (4.2.18), (4.2.20) and (4.2.14). To deal with the remaining two derivatives, we note that by (4.1.2)–(4.1.3), there holds

$$\left|\frac{d}{dt}\left(\int_0^1 v_x^2 dx\right)\right| = 2\left|\int_0^1 v_x v_{xt} dx\right| = 2\left|\int_0^1 v_t v_{xx} dx\right|$$

$$\leq C_1\int_0^1 (\theta_x^2 + \theta^2 u_x^2 u_x^2 v_x^2 + v_{xx}^2)\, dx,$$

$$\left|\frac{d}{dt}\left(\int_0^1 u_x^2 dx\right)\right| = 2\left|\int_0^1 u_x u_{xt} dx\right| = 2\left|\int_0^1 u_x v_{xx} dx\right| \leq C_1\int_0^1 (u_x^2 + v_{xx}^2)dx.$$

From (4.2.14), the proof of (4.2.4) is complete.

Note that with the boundary condition (4.1.8), we have

$$\left|\theta(x,t) - \int_0^1 \theta dx\right| \leq \left(\int_0^1 \theta_x^2 dx\right)^{1/2}$$

which, along with (4.2.14), gives

$$\theta(x,t) \leq C_1, \quad \forall (x,t) \in [0,1] \times [0,+\infty). \tag{4.2.23}$$

Defining $w(x,t) = \theta^{-1}(x,t)$, by (4.1.4), we get

$$w_t - \kappa \left(\frac{w_x}{u} \right)_x = \frac{w}{u^p} v_x - \mu \frac{w^2}{u} v_x^2 - 2\kappa \frac{w_x^2}{wu} - \delta w^2 f \leq \frac{w}{u^p} v_x - \mu \frac{w^2}{u} v_x^2. \tag{4.2.24}$$

Now define

$$w_+(x,t) = \max\{w(x,t) - \bar{\theta}^{-1}, 0\}.$$

Fix a natural number N and note that multiplying (4.2.24) by a nonnegative factor w_+^{N-1}, and by Young's inequality, we get

$$(w_+)_t w_+^{N-1} - \kappa \left(\frac{(w_+)_x}{u} \right)_x w_+^{N-1} \leq \frac{v_x}{u^p} w w_+^{N-1} - \mu \frac{v_x^2}{u} w^2 w_+^{N-1} \tag{4.2.25}$$

$$\leq C_1 \left(|v_x| w w_+^{(N-1)/2} \right)^2 + C_1 \left(w_+^{(N-1)/2} \right)^2 - \mu \frac{v_x^2}{u} w^2 w_+^{N-1} \leq C_1 w_+^{N-1}.$$

Integrating by parts and recalling the initial conditions we have

$$\int_0^1 \left(\frac{(w_+)_x}{u} \right)_x w_+^{N-1} dx = \frac{(w_+)_x}{u} w_+^{N-1} \Big|_0^1 - (N-1) \int_0^1 \frac{(w_+)_x^2}{u} w_+^{N-2} dx \leq 0$$

and thus, integrating (4.2.25) in space, by Hölder's inequality, we obtain

$$\frac{1}{N} \frac{d}{dt} \left(\int_0^1 w_+^N dx \right) \leq C_1 \int_0^1 w_+^{N-1} dx \leq C_1 \left(\int_0^1 w_+^N dx \right)^{(N-1)/N}.$$

Hence, for every N, there holds

$$\frac{d}{dt} \left(\int_0^1 w_+^N dx \right)^{1/N} = \frac{1}{N} \left(\int_0^1 w_+^N dx \right)^{1/(N-1)} \frac{d}{dt} \left(\int_0^1 w_+^N dx \right) \leq C_1$$

and we see that the following bound is true for every large number N and every $t \in [0,T]$:

$$\left(\int_0^1 w_+^N dx \right)^{1/N} \leq C_1(1+T). \tag{4.2.26}$$

Since the constant C_1 in (4.2.26) is independent of N, we have

$$\max_{t \in [0,T]} \max_{x \in [0,1]} w_+(x,t) \leq C_1(1+T)$$

which, along with (4.2.23), yields (4.2.1).

From (4.1.8) we infer

$$|v(x,t)| \leq \int_0^1 |v_x| dx \leq \left(\int_0^1 v_x^2 dx \right)^{1/2}. \tag{4.2.27}$$

Also,

$$|u(x,t) - 1| = |u(x,t) - \int_0^1 u dx| \leq \left(\int_0^1 u_x^2 dx \right)^{1/2}. \tag{4.2.28}$$

Thus, in view of (4.2.4), we have

$$\lim_{t \to +\infty} \max_{x \in [0,1]} (|v(x,t)| + |u(x,t) - 1|) = 0. \tag{4.2.29}$$

By (4.1.5) and (4.1.8), there holds

$$\frac{d}{dt} \left(\int_0^1 z dx \right) = - \int_0^1 f dx \leq 0. \tag{4.2.30}$$

On the other hand, by (4.1.5) and Young's inequality, we get

$$0 = \frac{d}{dt} \left(\int_0^1 z dx \right) + \int_0^1 f dx \geq \frac{d}{dt} \left(\int_0^1 z dx \right) + C_1^{-1} \left(\int_0^1 z dx \right)^m$$

and thus for large time t, there holds

$$\int_0^1 z(x,t) dx \leq \left\{ \begin{array}{ll} C_1 e^{-\lambda t}, & m = 1, \\ C_1 t^{-1/m-1}, & m > 1. \end{array} \right. \tag{4.2.31}$$

Since

$$\left| z(x,t) - \int_0^1 z(x,t) dx \right| \leq \left(\int_0^1 z_x^2 dx \right)^{1/2},$$

by (4.2.4), we receive

$$\lim_{t \to +\infty} \max_{x \in [0,1]} z(x,t) = 0. \tag{4.2.32}$$

Finally, we note that the quantity $\int_0^1 (\theta + v^2/2 + \delta z) dx$ is a constant in time, and thus

$$|\theta - \bar\theta| \leq \left| \theta(x,t) - \int_0^1 \theta dx \right| + \frac{1}{2} \int_0^1 v^2(x,t) dx + \delta \int_0^1 z(x,t) dx, \tag{4.2.33}$$

which, together with (4.2.33), (4.2.29) and (4.2.32), gives

$$\lim_{t \to +\infty} \max_{x \in [0,1]} |\theta(x,t) - \bar\theta| = 0. \tag{4.2.34}$$

Combining (4.2.29), (4.2.32) and (4.2.34), we obtain (4.2.5). The proof is complete. □

Lemma 4.2.2. *Under the assumptions of Theorem 4.1.1, the following estimate holds for any $t > 0$:*

$$\int_0^t (\|v_t\|^2 + \|\theta_t\|^2 + \|z_t\|^2 + \|f\|^2 + \|f_x\|^2 + \|f_t\|^2)(\tau)\, d\tau \le C_1. \qquad (4.2.35)$$

Proof. Since

$$0 < \exp\left(\frac{\theta - 1}{\varepsilon\theta}\right) \le C, \quad \forall \theta > 0,$$

it is easy to see that

$$\exp\left(\frac{\theta - 1}{\varepsilon\theta}\right)^2 \le C.$$

Combining this with (4.1.6) and (4.2.1)–(4.2.3), we have

$$\int_0^t \|f(\tau)\|^2\, d\tau = \int_0^t \int_0^1 f^2(x,\tau)\, dx d\tau \le C \int_0^t \int_0^1 f(x,\tau)\, dx d\tau \le C_1. \quad (4.2.36)$$

We easily infer from (4.1.3)–(4.1.6) and Lemma 4.2.1 that

$$\|v_t(t)\| \le C_1(\|u_x(t)\| + \|\theta_x(t)\| + \|v_x(t)\|_{L^\infty}\|u_x(t)\| + \|v_{xx}(t)\|)$$
$$\le C_2(\|u_x(t)\| + \|v_x(t)\|_{H^1} + \|\theta_x(t)\|), \qquad (4.2.37)$$
$$\|\theta_t(t)\| \le C_1(\|v_x(t)\| + \|v_x(t)\|_{L^\infty}\|v_x(t)\| + \|u_x(t)\|\|\theta_x(t)\|_{L^\infty}$$
$$+ \|\theta_{xx}(t)\| + \|f(t)\|)$$
$$\le C_1(\|v_x(t)\|_{H^1} + \|\theta_x(t)\|_{H^1} + \|f(t)\|), \qquad (4.2.38)$$
$$\|z_t(t)\| \le C_1(\|z_{xx}(t)\| + \|u_x(t)\|\|z_x(t)\|_{L^\infty} + \|f(t)\|)$$
$$\le C_1(\|z_x(t)\|_{H^1} + \|f(t)\|), \qquad (4.2.39)$$
$$\|f_x(t)\| \le C_1(\|u_x(t)\| + \|\theta_x(t)\| + \|z_x(t)\|), \qquad (4.2.40)$$
$$\|f_t(t)\| \le C_1(\|v_x(t)\| + \|\theta_t(t)\| + \|z_t(t)\|). \qquad (4.2.41)$$

Utilizing (4.2.3) and (4.2.36)–(4.2.41), we obtain (4.2.35). $\square$

Lemma 4.2.3. *Under the assumptions of Theorem 4.1.1, the following estimates hold for any $t > 0$:*

$$\|v_t(t)\|^2 + \|\theta_t(t)\|^2 + \|z_t(t)\|^2 + \|f_x(t)\|^2 + \|f_t(t)\|^2$$
$$+ \int_0^t (\|v_{tx}\|^2 + \|\theta_{tx}\|^2 + \|z_{tx}\|^2 + \|f_{tx}\|^2)(\tau)\, d\tau \le C_2, \qquad (4.2.42)$$

$$\|u_{xx}(t)\|^2 + \int_0^t \|u_{xx}(\tau)\|^2\, d\tau \le C_2, \qquad (4.2.43)$$

$$\|v_{xx}(t)\|^2 + \|\theta_{xx}(t)\|^2 + \|z_{xx}(t)\|^2$$
$$+ \int_0^t (\|v_{xxx}\|^2 + \|\theta_{xxx}\|^2 + \|z_{xxx}\|^2)(\tau)\, d\tau \le C_2. \qquad (4.2.44)$$

Proof. Since the proof is similar to that of Lemma 3.2.5 with a little modification, we only sketch the proof here. Only the estimates on θ are different due to the involvement of f.

Differentiating (4.1.3) with respect to t, multiplying the resulting equation by v_t in $L^2(0,1)$, performing an integration by parts, we obtain

$$\frac{1}{2}\frac{d}{dt}\|v_t(t)\|^2 = -\int_0^1 \left(-\frac{\theta}{u^p} + \mu\frac{v_x}{u}\right)_t v_{tx}\,dx$$

$$\leq -\mu\int_0^1 \frac{v_{tx}^2}{u}\,dx + C_1(\|v_x(t)\| + \|\theta_t(t)\| + \|v_x(t)\|_{L^4}^2)\|v_{tx}(t)\|$$

$$\leq -C_1^{-1}\|v_{tx}(t)\|^2 + C_1(\|v_x(t)\|^2 + \|\theta_t(t)\|^2 + \|v_{xx}(t)\|^2) \quad (4.2.45)$$

which, together with Lemmas 4.2.1–4.2.2, yields

$$\|v_t(t)\|^2 + \int_0^t \|v_{tx}(\tau)\|^2\,d\tau \leq C_2, \quad \forall\, t > 0. \quad (4.2.46)$$

On the other hand, using (4.1.3) and (4.2.46), Lemmas 4.2.1–4.2.2, Sobolev's embedding theorem and Young's inequality, we have

$$\|v_{xx}(t)\| \leq C_1(\|u_x(t)\|\|v_x(t)\|_{L^\infty} + \|\theta_x(t)\| + \|v_t\| + \|u_x(t)\|)$$

$$\leq \frac{1}{2}\|v_{xx}(t)\| + C_1(\|u_x(t)\| + \|v_x(t)\| + \|\theta_x(t)\| + \|v_t(t)\|)$$

which leads to

$$\|v_{xx}(t)\| \leq C_2, \quad \|v_x(t)\|_{L^\infty} \leq C_2, \quad \forall\, t > 0. \quad (4.2.47)$$

Similarly,

$$\|\theta_{xx}(t)\| \leq C_1(\|v_x(t)\|_{H^1} + \|\theta_x(t)\| + \|f(t)\| + \|\theta_t(t)\|), \quad (4.2.48)$$

$$\|z_{xx}(t)\| \leq C_1(\|f(t)\| + \|z_x(t)\| + \|z_t(t)\|). \quad (4.2.49)$$

Similarly, differentiating (4.1.4) with respect to t, multiplying the resultant by θ_t and integrating by parts, we deduce

$$\frac{1}{2}\frac{d}{dt}\|\theta_t(t)\|^2 + C_1^{-1}\|\theta_{tx}(t)\|^2$$

$$\leq (2C_1)^{-1}\|\theta_{tx}(t)\|^2 + C_1\Big(\|\theta_x(t)\|^2 + \|v_x(t)\|_{H^1}^2 + \|\theta_t(t)\|^2$$

$$+ \|\theta_t(t)\|^2\|v_x(t)\|_{L^\infty}^2 + \|v_{tx}(t)\|^2 + \|f_t(t)\|^2\Big) \quad (4.2.50)$$

which, combined with (4.2.46) and Lemmas 4.2.1–4.2.2, implies

$$\|\theta_t(t)\|^2 + \int_0^t \|\theta_{tx}(\tau)\|^2\,d\tau \leq C_2, \quad \forall\, t > 0. \quad (4.2.51)$$

By (4.2.48) and (4.2.51), we easily get

$$\|\theta_{xx}(t)\| \leq C_2, \quad \|\theta_x(t)\|_{L^\infty} \leq C_2, \quad \forall\, t > 0. \quad (4.2.52)$$

Similarly to (4.2.46) and (4.2.51), by equation (4.1.5), we have

$$\frac{1}{2}\frac{d}{dt}\|z_t(t)\|^2 + C_1^{-1}\|z_{tx}(t)\|^2 \tag{4.2.53}$$
$$\leq (2C_1)^{-1}\|z_{tx}(t)\|^2 + C_1(\|v_x(t)\|_{L^\infty}^2\|z_x(t)\|^2 + \|z_t(t)\|^2 + \|f_t(t)\|^2)$$

which, combined with (4.2.47), (4.2.49) and Lemmas 4.2.1–4.2.2, implies

$$\|z_t(t)\|^2 + \int_0^t \|z_{tx}(\tau)\|^2\, d\tau \leq C_2, \quad \forall\, t > 0, \tag{4.2.54}$$

$$\|z_{xx}(t)\| + \|z_x(t)\|_{L^\infty} \leq C_2, \quad \forall\, t > 0. \tag{4.2.55}$$

Differentiating (4.1.3) with respect to x, using (4.1.2) ($u_{txx} = v_{xxx}$), we see that

$$\mu\frac{\partial}{\partial t}\left(\frac{u_{xx}}{u}\right) - \mathcal{P}_u u_{xx} = v_{tx} + E(x,t) \tag{4.2.56}$$

where

$$E(x,t) = \mathcal{P}_{uu}u_x^2 + 2\mathcal{P}_{\theta u}\theta_x u_x + \mathcal{P}_{\theta\theta}\theta_x^2 + \mathcal{P}_\theta\theta_{xx} - 2\mu\frac{v_x u_x^2}{u^3} + 2\mu\frac{u_x v_{xx}}{u^2}$$
$$= p(p+1)\frac{\theta}{u^{p+2}}u_x^2 - 2p\frac{\theta_x u_x}{u^{p+1}} + \frac{1}{u^{p+1}}\theta_{xx} - 2\mu\left(\frac{v_{xx}u_x}{u^2} - \frac{v_x u_x^2}{u^3}\right).$$

Multiplying (4.2.56) by u_{xx}/u, and by Young's inequality, by (4.2.46)–(4.2.47), (4.2.51)–(4.2.52), (4.2.54)–(4.2.55) and Lemmas 4.2.1–4.2.2, we can deduce that

$$\frac{d}{dt}\left\|\frac{u_{xx}}{u}(t)\right\|^2 + C_1^{-1}\left\|\frac{u_{xx}}{u}(t)\right\|^2$$
$$\leq \frac{1}{4C_1}\left\|\frac{u_{xx}}{u}(t)\right\|^2 + C_1\left(\|\theta_x(t)\|_{L^\infty}^2\|u_x(t)\|^2 + \|u_x(t)\|_{L^4}^4 + \|v_{tx}(t)\|^2\right.$$
$$\left. + \|u_x(t)\|_{L^\infty}^2\|v_{xx}(t)\|^2 + \|\theta_{xx}(t)\|^2 + \|v_x(t)\|_{L^\infty}^2\|u_x(t)\|_{L^4}^4\right)$$
$$\leq \frac{1}{2C_1}\left\|\frac{u_{xx}}{u}\right\|^2 + C_2(\|\theta_x(t)\|^2 + \|\theta_{xx}(t)\|^2 + \|u_x(t)\|^2 + \|v_{tx}(t)\|^2) \tag{4.2.57}$$

which, combined with Lemma 4.2.1 and (4.2.46), gives (4.2.43).

Differentiating (4.1.3), (4.1.4) and (4.1.5) with respect to x, respectively, and using (4.2.47), (4.2.52), (4.2.55) and Lemmas 4.2.1–4.2.2 to get

$$\|v_{tx}(t)\| \leq C_2(\|v_x(t)\|_{H^2} + \|u_x(t)\|_{H^1} + \|\theta_x(t)\|_{H^1}) \tag{4.2.58}$$
$$\|\theta_{tx}(t)\| \leq C_2(\|\theta_x(t)\|_{H^2} + \|u_x(t)\|_{H^1} + \|v_x(t)\|_{H^1} + \|f_x(t)\|), \tag{4.2.59}$$
$$\|z_{tx}(t)\| \leq C_1(\|f_x(t)\| + \|z_{xxx}(t)\| + \|u_x(t)\|_{L^\infty}\|z_{xx}(t)\|$$
$$+ \|z_x(t)\|_{L^\infty}\|u_{xx}(t)\| + \|u_x(t)\|_{L^\infty}^2\|z_x(t)\|)$$
$$\leq C_2(\|z_x(t)\|_{H^2} + \|f_x(t)\| + \|u_x(t)\|_{H^1}) \tag{4.2.60}$$

or

$$\|v_{xxx}(t)\| \leq C_2(\|v_x(t)\|_{H^1} + \|u_x(t)\|_{H^1} + \|\theta_x(t)\|_{H^1} + \|v_{tx}(t)\|), \qquad (4.2.61)$$

$$\|\theta_{xxx}(t)\| \leq C_2(\|\theta_x(t)\|_{H^1} + \|u_x(t)\|_{H^1} + \|v_x(t)\|_{H^1} + \|f_x(t)\| + \|\theta_{tx}(t)\|), \qquad (4.2.62)$$

$$\|z_{xxx}(t)\| \leq C_2(\|z_x(t)\|_{H^1} + \|f_x(t)\| + \|u_x(t)\|_{H^1} + \|z_{tx}(t)\|). \qquad (4.2.63)$$

We derive from (4.1.6), (4.2.47), (4.2.52) and (4.2.55) that

$$\|f_{tx}(t)\| \leq C_1(\|v_x(t)\|_{H^1} + \|z_t(t)\| + \|\theta_t(t)\| + \|\theta_{tx}(t)\| + \|z_{tx}(t)\|). \qquad (4.2.64)$$

By (4.2.40)–(4.2.41), (4.2.46)–(4.2.47), (4.2.51)–(4.2.52), (4.2.54)–(4.2.55), (4.2.64) and Lemmas 4.2.1–4.2.2, we obtain (4.2.42), and from (4.2.42)–(4.2.43) and (4.2.61)–(4.2.63), we conclude (4.2.44). The proof is complete. $\qquad \square$

4.3 Exponential Stability in H^2

In this section, we shall establish the exponential stability of solutions in H_+^2.

Let $\rho = \frac{1}{u}$, and we easily get that the specific entropy (4.2.6) satisfies

$$\frac{\partial \eta}{\partial \rho} = \begin{cases} -\mathcal{P}_\theta/\rho^2, & p = 1, \\ 1/\mathcal{P}_\theta, & p > 1, \end{cases}, \qquad \frac{\partial \eta}{\partial \theta} = \frac{e_\theta}{\theta} = \frac{1}{\theta} \qquad (4.3.1)$$

with $\mathcal{P} = \rho^p \theta$.

We consider the transform

$$\mathcal{A} : (\rho, \theta) \in \mathcal{D}_{\rho,\theta} = \{(\rho, \theta) : \rho > 0, \ \theta > 0\} \to (u, \eta) \in \mathcal{A}\mathcal{D}_{\rho,\theta},$$

where $u = 1/\rho$ and $\eta = \eta(1/\rho, \theta)$. Owing to the Jacobian

$$\frac{\partial(u, \eta)}{\partial(\rho, \theta)} = -\frac{e_\theta}{\rho^2 \theta} = -\frac{1}{\rho^2 \theta} < 0, \qquad (4.3.2)$$

there is a unique inverse function $(u, \eta) \in \mathcal{A}\mathcal{D}_{\rho,\theta}$. Thus the function e, p can be also regarded as the smooth functions of (u, η). We write

$$e = e(u, \eta) :\equiv e(u, \theta(u, \eta)) = e(1/\rho, \theta),$$

$$\mathcal{P} = \mathcal{P}(u, \eta) :\equiv \mathcal{P}(u, \theta(u, \eta)) = \mathcal{P}(1/\rho, \theta).$$

Let

$$\mathcal{E}(u, v, \eta, z) = \frac{v^2}{2} + \delta z + e(u, \eta) - e(\bar{u}, \bar{\eta}) - \frac{\partial e}{\partial u}(\bar{u}, \bar{\eta})(u - \bar{u}) - \frac{\partial e}{\partial \eta}(\bar{u}, \bar{\eta})(\eta - \bar{\eta}), \qquad (4.3.3)$$

where $e(u, \eta) = e(u, \theta) = \theta$, $\bar{u} = \int_0^1 u_0 \, dx = 1$ and $\bar{\theta} > 0$ is determined by

$$e(\bar{u}, \bar{\theta}) = e(\bar{u}, \bar{\eta}) = \int_0^1 \left(\frac{1}{2} v_0^2 + e(u_0, \theta_0) + \delta z_0 \right) dx \qquad (4.3.4)$$

and

$$\bar{\eta} = \eta(\bar{u}, \bar{\theta}) = \eta(1, \bar{\theta}). \qquad (4.3.5)$$

Lemma 4.3.1. *Under assumptions of Theorem 4.1.1, there holds*

$$\frac{v^2}{2} + \delta z + C_1^{-1} \left(|u - 1|^2 + |\eta - \bar{\eta}|^2 \right)$$

$$\leq \mathcal{E}(u, v, \eta, z) \leq \frac{v^2}{2} + \delta z + C_1 \left(|u - 1|^2 + |\eta - \bar{\eta}|^2 \right). \tag{4.3.6}$$

Proof. The proof is similar to that of Lemma 3.2.3.   □

Lemma 4.3.2. *Under assumptions of Theorem 4.1.1, there are positive constants $C_1 > 0$ and $\gamma_1' = \gamma_1'(C_1) < \gamma_0/2$ such that for any fixed $\gamma \in (0, \gamma_1']$, there holds for any $t > 0$,*

$$e^{\gamma t} \Big(\|v(t)\|^2 + \|z(t)\|_{L^1[0,1]} + \|z(t)\|^2 + \|u(t) - 1\|^2$$

$$+ \|\eta(t) - \bar{\eta}\|^2 + \|\rho_x(t)\|^2 + \|u_x(t)\|^2 \Big)$$

$$+ \int_0^t e^{\gamma t} \Big(\|z\|_{H^1}^2 + \|\theta_x\|^2 + \|v_x\|^2$$

$$+ \int_0^1 f(u, \theta, z) \, dx + \|u_x\|^2 + \|\rho_x\|^2 \Big)(s) \, ds \leq C_1. \tag{4.3.7}$$

Proof. By (4.1.1)–(4.1.6) and (4.2.6), we easily infer that the following equations hold:

$$\left(\theta + \frac{v^2}{2} + \delta z \right)_t = \left(\sigma v + \frac{k\theta_x}{u} + \frac{\delta d}{u^2} z_x \right)_x, \tag{4.3.8}$$

$$\eta_t = (k\rho\theta_x/\theta)_x + k\rho \left(\theta_x/\theta \right)^2 + \frac{\mu\rho v_x^2}{\theta} + \frac{\delta f}{\theta}. \tag{4.3.9}$$

From (4.1.1)–(4.1.6), (4.3.8)–(4.3.9), we can infer

$$\mathcal{E}_t + \bar{\theta} \left[k\rho(\theta_x/\theta)^2 + \mu\rho v_x^2/\theta + \frac{\delta f}{\theta} \right]$$

$$= \left[\mu\rho v v_x - v(\mathcal{P} - \bar{\mathcal{P}}) \right]_x + \left[(1 - \bar{\theta}/\theta)k\rho\theta_x \right]_x + \delta \left(d\rho^2 z_x \right)_x \tag{4.3.10}$$

with $\bar{\mathcal{P}} = \mathcal{P}(\bar{u}, \bar{\theta})$ and

$$\left(\mu^2 \frac{(\rho_x/\rho)^2}{2} + \mu \frac{\rho_x v}{\rho} \right)_t + \mu \frac{\mathcal{P}_\rho \rho_x^2}{\rho} = -\mu \frac{\mathcal{P}_\theta \rho_x \theta_x}{\rho} - \mu(\rho v v_x)_x + \mu\rho v_x^2. \tag{4.3.11}$$

For any $\beta, \gamma > 0$, let

$$G(t) = e^{\gamma t} \left[\mathcal{E} + \beta \left(\mu^2 (\rho_x/\rho)^2/2 + \mu\rho_x v/\rho \right) \right]. \tag{4.3.12}$$

Multiplying (4.3.10) and (4.3.11) by $e^{\gamma t}$, $\beta e^{\gamma t}$ respectively, and then adding the resulting equations, we conclude

$$
\begin{aligned}
G'(t) + e^{\gamma t} &\left[\frac{\bar{\theta}}{\theta} \left(k\rho\theta_x^2/\theta + \mu\rho v_x^2 + \delta f(u,\theta,z) \right) \right] \\
&+ \beta e^{\gamma t} \left[\mu \mathcal{P}_\rho \rho_x^2/\rho + \mu \mathcal{P}_\theta \rho_x \theta_x/\rho - \mu\rho v_x^2 \right] \\
&= \gamma e^{\gamma t} \left[\mathcal{E} + \beta \left(\mu^2 (\rho_x/\rho)^2/2 + \mu\rho_x v/\rho \right) \right] \\
&+ e^{\gamma t} \left[\mu(1-\beta)\rho v v_x + k \left(1 - \frac{\bar{\theta}}{\theta} \right) \rho\theta_x + \delta d\rho^2 z_x - (\mathcal{P} - \bar{\mathcal{P}})v \right]_x. \quad (4.3.13)
\end{aligned}
$$

By the mean value theorem and Lemmas 4.2.1–4.2.3, we get

$$
\begin{aligned}
\|\eta(t) - \bar{\eta}\| = \|\eta(u,\theta) - \eta(\bar{u},\bar{\theta})\| &\leq C_1(\|u(t) - \bar{u}\| + \|\theta(t) - \bar{\theta}\|) \\
&\leq C_1(\|u_x(t)\| + \|\theta(t) - \bar{\theta}\|). \quad (4.3.14)
\end{aligned}
$$

Noting that

$$
e(\bar{u},\bar{\theta}) = \int_0^1 \left(\frac{1}{2}v_0^2 + \theta_0 + \delta z_0 \right) dx = \int_0^1 \left(\frac{1}{2}v^2 + \theta + \delta z \right) dx, \quad (4.3.15)
$$

we infer, by the mean value theorem,

$$
\begin{aligned}
\|\theta(t) - \bar{\theta}\| &\leq C_1(\|u(t) - \bar{u}\| + \|e(u,\theta) - e(\bar{u},\bar{\theta})\|) \\
&\leq C_1(\|u(t) - \bar{u}\| + \|e(u,\theta) - \overline{e(u,\theta)}\| + \|v(t)\| + \|z(t)\|_{L^1[0,1]}) \\
&\leq C_1(\|u_x(t)\| + \|e_x(t)\| + \|v_x(t)\| + \|z(t)\|) \\
&\leq C_1(\|u_x(t)\| + \|\theta_x(t)\| + \|v_x(t)\| + \|z(t)\|) \quad (4.3.16)
\end{aligned}
$$

which, along with (4.3.14) and the Poincaré inequality, gives

$$
\|\eta - \bar{\eta}\| \leq C_1(\|u_x(t)\| + \|\theta_x(t)\| + \|v_x(t)\| + \|z(t)\|) \quad (4.3.17)
$$

where $\overline{e(u,\theta)} = \int_0^1 e(u,\theta)\,dx$.

The following estimate is crucial to our next argument, which implies that $\|z\|$ decays exponentially. Multiplying (4.1.5) by z, integrating the resulting equation in $L^2(0,1)$, and using Lemmas 4.2.1–4.2.3, we infer

$$
\frac{1}{2}\frac{d}{dt}\|z(t)\|^2 + \int_0^1 \frac{d}{u^2}z_x^2\,dx + \int_0^1 f(u,\theta,z)z\,dx = 0
$$

which implies

$$
\frac{d}{dt}\|z(t)\|^2 + \gamma_0\|z_x(t)\|^2 + \gamma_0\|z(t)\|^2 \leq 0
$$

i.e.,

$$
e^{\gamma_0 t}\|z(t)\|^2 + \int_0^t e^{\gamma_0 s}\|z_x(s)\|^2\,ds \leq C_1, \quad \forall t > 0 \quad (4.3.18)
$$

with a positive constant $\gamma_0 = \gamma_0(C_1) > 0$.

Integrating (4.3.13) with respect to x and t, using (4.3.17) and (4.3.18), we conclude

$$\int_0^1 G(t)dx + \int_0^t e^{\gamma s} \int_0^1 \left[\frac{\bar{\theta}}{\theta} \left(k\rho\theta_x^2/\theta + \mu\rho v_x^2 + \delta f(u,\theta,z) \right) \right] (x,s)\,dxds$$

$$+ \beta \int_0^t \int_0^1 e^{\gamma s} \left[\mu \mathcal{P}_\rho \rho_x^2/\rho + \mu \mathcal{P}_\theta \rho_x \theta_x/\rho - \mu\rho v_x^2 \right] (x,s)\,dxds$$

$$= \int_0^1 G(0)\,dx + \gamma \int_0^t \int_0^1 e^{\gamma s} \left[\mathcal{E} + \beta \left(\mu^2 (\rho_x/\rho)^2/2 + \mu\rho_x v/\rho \right) \right] (x,s)\,dxds$$

$$\le C_1 + C_1\gamma \int_0^t e^{\gamma s} \left(\|v\|^2 + \|z\|_{L^1(0,1)} + \|u - \bar{u}\|^2 + \|\eta - \bar{\eta}\|^2 + \|\rho_x\|^2 \right)(s)\,ds$$

$$\le C_1 + C_1\gamma \int_0^t e^{\gamma s} \left(\|v_x\|^2 + \|z\|_{L^1(0,1)} + \|u_x\|^2 + \|\theta_x\|^2 + \|z\|^2 + \|\rho_x\|^2 \right)(s)\,ds$$

$$\le C_1 + C_1\gamma \int_0^t e^{\gamma s} \left(\|v_x\|^2 + \|z\| + \|u_x\|^2 + \|\theta_x\|^2 + \|z\|^2 + \|\rho_x\|^2 \right)(s)\,ds$$

$$\le C_1(\gamma) + C_1\gamma \int_0^t e^{\gamma s} \left(\|v_x\|^2 + \|u_x\|^2 + \|\rho_x\|^2 + \|\theta_x\|^2 \right)(s)\,ds \qquad (4.3.19)$$

if we choose $0 < \gamma < \gamma_0/2$.

From Lemma 4.3.1 and Lemmas 4.2.1–4.2.3, we easily infer

$$\int_0^1 G(t)\,dx \ge e^{\gamma t} \Big\{ C_1^{-1} (\|v(t)\|^2 + \|z(t)\|_{L^1(0,1)} + \|u(t) - 1\|^2 + \|\eta(t) - \bar{\eta}\|^2)$$

$$+ \beta \int_0^1 \left(\mu^2 (\rho_x/\rho)^2 + \mu v \rho_x/\rho \right)\,dx \Big\}$$

$$\ge e^{\gamma t} \Big\{ C_1^{-1} (\|v(t)\|^2 + \|z(t)\|_{L^1(0,1)} + \|u(t) - 1\|^2 + \|\eta(t) - \bar{\eta}\|^2)$$

$$+ \beta C_1^{-1} \|\rho_x(t)\|^2 - C_1\beta \|v(t)\|^2 \Big\}$$

$$\ge C_1^{-1} e^{\gamma t} \Big(\|v(t)\|^2 + \|z(t)\|_{L^1(0,1)} + \|u(t) - 1\|^2 + \|\eta(t)$$

$$- \bar{\eta}\|^2 + \beta \|\rho_x(t)\|^2 \Big) \qquad (4.3.20)$$

if β is small enough.

The Young inequality yields

$$\frac{\mathcal{P}_\theta \rho_x \theta_x}{\rho} \ge -\frac{1}{2} \mathcal{P}_\rho \frac{\rho_x^2}{\rho} - C_1 \theta_x^2 \qquad (4.3.21)$$

with $\mathcal{P}_\rho = p\rho^{p-1}\theta > 0$. Thus it follows from (4.3.16), (4.3.19)–(4.3.21) that for $\beta = \gamma^{1/2}$, there are constants $C_1 > 0$ and $0 < \gamma_1' = \gamma_1'(C_1) < \gamma_0/2$ such that for $\gamma \in (0, \gamma'_1]$,

$$e^{\gamma t} \left(\|v(t)\|^2 + \|z(t)\|_{L^1(0,1)} + \|u(t) - 1\|^2 + \|\eta(t) - \bar{\eta}\|^2 \right.$$
$$\left. + \gamma^{\frac{1}{2}} \|\rho_x(t)\|^2 + \gamma^{\frac{1}{2}} \|u_x(t)\|^2 \right)$$
$$+ \int_0^t e^{\gamma s} (\|v_x\|^2 + \|\theta_x\|^2 + C_1^{-1} \int_0^1 f(u, \theta, z)\, dx + \gamma^{\frac{1}{2}} \|\rho_x\|^2 + \gamma^{\frac{1}{2}} \|u_x\|^2)(s)\, ds$$
$$\leq C_1(\gamma) + C_1 \gamma \int_0^t e^{\gamma s} (\|v_x\|^2 + \|u_x\|^2 + \|\rho_x\|^2 + \|\theta_x\|^2)(s)\, ds$$

which, by choosing $0 < \gamma < \gamma'_1$ small enough and using (4.3.18), yields (4.3.7). $\qquad \square$

Lemma 4.3.3. *Under assumptions of Theorem 4.1.1, there are positive constants $C_1 > 0$ and $\gamma_1 = \gamma_1(C_1) \leq \gamma'_1$ such that for any fixed $\gamma \in (0, \gamma_1]$, the following estimate holds,*

$$e^{\gamma t} (\|v_x(t)\|^2 + \|\theta_x(t)\|^2 + \|z_x(t)\|^2) \qquad (4.3.22)$$
$$+ \int_0^t e^{\gamma s} \left(\|v_{xx}\|^2 + \|\theta_{xx}\|^2 + \|z_{xx}\|^2 + \|v_t\|^2 + \|\theta_t\|^2 + \|z_t\|^2 \right)(s)\, ds \leq C_1.$$

Proof. Multiplying (4.1.3) by $e^{\gamma t} v_{xx}$, respectively, using Lemmas 4.2.1–4.2.3, we infer from

$$\frac{1}{2} \frac{d}{dt} (e^{\gamma t} \|v_x\|^2) + \mu e^{\gamma t} \int_0^1 \frac{v_{xx}^2}{u}\, dx$$
$$= e^{\gamma t} \int_0^1 \left(P_x + \frac{\mu v_x u_x}{u^2} \right) dx + \frac{\gamma}{2} e^{\gamma t} \|v_x\|^2$$
$$\leq \frac{\gamma}{2} e^{\gamma t} \|v_x\|^2 + C_1 e^{\gamma t} (\|u_x\|^2 + \|\theta_x\|^2 + \|v_x\|_{L^\infty} \|u_x\|)$$
$$\leq \frac{\gamma}{2} e^{\gamma t} \|v_x\|^2 + + C_1 e^{\gamma t} (\|u_x\|^2 + \|\theta_x\|^2 + \|v_x\|^{\frac{1}{2}} \|v_{xx}\|^{\frac{1}{2}} \|u_x\|)$$
$$\leq C_1 \gamma e^{\gamma t} \|v_{xx}\|^2 + C_1 e^{\gamma t} (\|u_x\|^2 + \|\theta_x\|^2 + \|v_x\|^2),$$

i.e., for $\gamma > 0$ small enough,

$$\frac{1}{2} \frac{d}{dt} (e^{\gamma t} \|v_x\|^2) + (2C_1)^{-1} e^{\gamma t} \|v_{xx}\|^2$$
$$\leq C_1 \gamma^{-1} e^{\gamma t} (\|u_x\|^2 + \|\theta_x\|^2 + \|v_x\|^2). \qquad (4.3.23)$$

Multiplying (4.1.4) by $e^{\gamma t} \theta_{xx}$, using Lemmas 4.2.1–4.2.3, we infer from

$$\frac{1}{2} \frac{d}{dt} (e^{\gamma t} \|\theta_x(t)\|^2) + e^{\gamma t} \int_0^1 \frac{k \theta_{xx}^2}{u}\, dx$$
$$\leq \frac{\gamma}{2} e^{\gamma t} \|\theta_x\|^2 + C_1 e^{\gamma t} \left(\|\theta_x\|_{L^\infty} \|u_x\| + \|v_x\| + \|v_x\|_{L^\infty} + \int_0^1 f(u, \theta, z)\, dx \right) \|\theta_{xx}\|$$

$$\leq \frac{\gamma}{2}e^{\gamma t}\|\theta_x\|^2 + C_1 e^{\gamma t}\left(\|\theta_x\|^{\frac{1}{2}}\|\theta_{xx}\|^{\frac{3}{2}}\|u_x\|\right.$$

$$\left. + \|v_x\|\|\theta_{xx}\| + \|v_x\|^{\frac{1}{2}}\|v_{xx}\|^{\frac{1}{2}}\|\theta_{xx}\| + \int_0^1 f(u,\theta,z)\,dx\|\theta_{xx}\|\right)$$

$$\leq \frac{\gamma}{2}e^{\gamma t}\|\theta_{xx}\|^2 + \gamma e^{\gamma t}\|v_{xx}\|^2 + \frac{\gamma}{2}e^{\gamma t}\|\theta_{xx}\|^2 + C_1(\gamma)e^{\gamma t}\left[\|\theta_x\|^2 + \|v_x\|^2 + \|z\|^2\right],$$

i.e., for $\gamma > 0$ small enough,

$$\frac{1}{2}\frac{d}{dt}(e^{\gamma t}\|\theta_x\|^2) + \frac{1}{2C_1}e^{\gamma t}\|\theta_{xx}\|^2 \leq \gamma e^{\gamma t}(\|v_{xx}\|^2 + \|\theta_{xx}\|^2) + C_1 e^{\gamma t}\|z\|^2. \quad (4.3.24)$$

Similarly, multiplying (4.1.5) by $e^{\gamma t}z_{xx}$, using Lemmas 4.2.1–4.2.3, we conclude for $\gamma > 0$ small enough,

$$\frac{1}{2}\frac{d}{dt}(e^{\gamma t}\|z_x\|^2) + (2C_1)^{-1}e^{\gamma t}\|z_{xx}\|^2 \leq C_1\gamma^{-1}e^{\gamma t}(\|u_x\|^2 + \|z\|^2). \quad (4.3.25)$$

Adding (4.3.23), (4.3.24) and (4.3.25), we arrive at

$$\frac{1}{2}\frac{d}{dt}\left[e^{\gamma t}(\|v_x\|^2 + \|\theta_x\|^2 + \|z_x\|^2)\right] + \frac{1}{2C_1}e^{\gamma t}(\|v_{xx}\|^2 + \|\theta_{xx}\|^2 + \|z_{xx}\|^2)$$

$$\leq \gamma e^{\gamma t}(\|v_{xx}\|^2 + \|\theta_{xx}\|^2) + C_1\gamma^{-1}e^{\gamma t}(\|u_x\|^2 + \|v_x\|^2 + \|\theta_x\|^2 + \|z\|^2) \quad (4.3.26)$$

for $\gamma > 0$ small enough. This, along with Lemma 4.3.2, gives (4.3.22) by using (4.3.18) and (4.1.2)–(4.1.5) for $\gamma \in (0, \gamma_1']$ small enough. $\square$

Lemma 4.3.4. *Under assumptions of Theorem 4.1.1, there exists a positive constant $\gamma_2' = \gamma_2'(C_2) \leq \gamma_1$ such that for any fixed $\gamma \in (0, \gamma_2']$, the following estimate holds,*

$$e^{\gamma t}\left(\|v_t(t)\|^2 + \|\theta_t(t)\|^2 + \|z_t(t)\|^2 + \|v_{xx}(t)\|^2 + \|\theta_{xx}(t)\|^2 + \|z_{xx}(t)\|^2\right)$$

$$+ \int_0^t e^{\gamma \tau}(\|v_{tx}\|^2 + \|\theta_{tx}\|^2 + \|z_{tx}\|^2)(\tau)\,d\tau \leq C_2, \quad \forall t > 0. \quad (4.3.27)$$

Proof. Multiplying (4.2.45) by $e^{\gamma t}$ and integrating the resulting equation over $[0, t]$, using Young's inequality, we easily conclude

$$e^{\gamma t}\|v_t(t)\|^2 + \int_0^t e^{\gamma \tau}\|v_{tx}(\tau)\|^2\,d\tau \leq C_2 + \int_0^t e^{\gamma \tau}\left(\|v_t\|^2 + \|\theta_t\|^2 + \|v_x\|_{H^1}^2\right)(\tau)\,d\tau$$

which, combined with Lemmas 4.3.2–4.3.3 and Lemmas 4.2.1–4.2.3, implies that there exists a constant $\gamma_2' = \gamma_2'(C_2) \leq \gamma_1$ such that for any fixed $\gamma \in (0, \gamma_2']$

$$e^{\gamma t}(\|v_t(t)\|^2 + \|v_{xx}(t)\|^2) + \int_0^t e^{\gamma \tau}\|v_{tx}(\tau)\|^2\,d\tau \leq C_2, \quad \forall t > 0. \quad (4.3.28)$$

In the same manner, multiplying (4.2.50) and (4.2.53) by $e^{\gamma t}$, respectively, integrating the resulting equations over $[0, t]$ and using Lemmas 4.3.1–4.3.3 and (4.2.48)–(4.2.49), we infer that

$$e^{\gamma t}(\|\theta_t(t)\|^2 + \|z_t(t)\|^2 + \|\theta_{xx}(t)\|^2 + \|z_{xx}(t)\|^2) + \int_0^t e^{\gamma \tau}(\|\theta_{tx}\|^2 + \|z_{tx}\|^2)(\tau)\, d\tau \le C_2$$

which, together with (4.3.28), yields (4.3.27). The proof is complete. $\qquad\square$

Lemma 4.3.5. *Under assumptions of Theorem* 4.1.1, *there exists a positive constant* $\gamma_2 = \gamma_2(C_2) \le \gamma_2'$ *such that for any fixed* $\gamma \in (0, \gamma_2]$, *the following estimate holds,*

$$e^{\gamma t}\|u_{xx}(t)\|^2 + \int_0^t e^{\gamma \tau}(\|u_{xx}\|^2 + \|v_{xxx}\|^2 + \|\theta_{xxx}\|^2 + \|z_{xxx}\|^2)(\tau)\, d\tau \le C_2, \quad \forall t > 0.$$
$$(4.3.29)$$

Proof. Multiplying (4.2.57) by $e^{\gamma t}$ and choosing γ so small that

$$\gamma \le \gamma_2 = \min[\gamma_2', 1/(4C_1)],$$

and using Lemmas 4.3.1–4.3.4 and (4.2.61)–(4.2.63), we conclude that

$$e^{\gamma t}\|u_{xx}(t)\|^2 + \int_0^t e^{\gamma \tau}\|u_{xx}(\tau)\|^2\, d\tau$$
$$\le C_2 + C_1 \int_0^t e^{\gamma \tau}(\|\theta_{xx}\|^2 + \|u_x\|^2 + \|v_{tx}\|^2)(\tau)\, d\tau$$
$$\le C_2. \qquad\qquad (4.3.30)$$

Differentiating (4.1.3), (4.1.4) and (4.1.5) with respect to x, respectively, using Lemmas 4.3.1–4.3.4, and (4.3.30), we easily deduce

$$\int_0^t e^{\gamma \tau}(\|v_{xxx}\|^2 + \|\theta_{xxx}\|^2 + \|z_{xxx}\|^2)(\tau)\, d\tau \le C_2$$

which, along with (4.3.30), gives (4.3.29). The proof is complete. $\qquad\square$

Proof of Theorem 4.1.1. By Lemmas 4.2.1–4.2.3, Lemmas 4.3.1–4.3.5 and Sobolev's embedding theorem, we complete the proof of Theorem 4.1.1. $\qquad\square$

4.4   Proof of Theorem 4.1.2

In this section, we shall give the proof of Theorem 4.1.2 by a series of a priori estimates.

Lemma 4.4.1. *Under assumptions of Theorem 4.1.2, for any $(u_0, v_0, \theta_0, z_0) \in H^4_+$, we have*

$$\|v_{tx}(x,0)\| + \|\theta_{tx}(x,0)\| + \|z_{tx}(x,0)\| \le C_3, \tag{4.4.1}$$

$$\|v_{tt}(x,0)\| + \|\theta_{tt}(x,0)\| + \|z_{tt}(x,0)\| + \|v_{txx}(x,0)\|$$
$$+ \|\theta_{txx}(x,0)\| + \|z_{txx}(x,0)\| \le C_4, \tag{4.4.2}$$

$$\|v_{tt}(t)\|^2 + \int_0^t \|v_{ttx}(\tau)\|^2 \, d\tau \le C_4 + C_4 \int_0^t \|\theta_{txx}(\tau)\|^2 \, d\tau, \quad \forall t > 0, \tag{4.4.3}$$

$$\|\theta_{tt}(t)\|^2 + \int_0^t \|\theta_{ttx}(\tau)\|^2 \, d\tau$$

$$\le C_4 + C_2 \varepsilon^{-1} \int_0^t \|\theta_{txx}(\tau)\|^2 \, d\tau$$

$$+ C_1 \varepsilon \int_0^t (\|v_{ttx}\|^2 + \|v_{txx}\|^2 + \|z_{txx}\|^2)(\tau) \, d\tau, \quad \forall t > 0, \tag{4.4.4}$$

$$\|z_{tt}(t)\|^2 + \int_0^t \|z_{ttx}(\tau)\|^2 \, d\tau \le C_4 + C_4 \int_0^t (\|\theta_{txx}\|^2 + \|z_{txx}\|^2)(\tau) \, d\tau, \quad \forall t > 0 \tag{4.4.5}$$

for ε small enough.

Proof. By virtue of (4.2.58)–(4.2.60) and (4.2.40), we conclude (4.4.1).

Differentiating (4.1.3), (4.1.4) and (4.1.5) with respect to x twice, respectively, and using Lemmas 4.2.1–4.2.3 to get

$$\|v_{txx}(t)\| \le C_2(\|u_x(t)\|_{H^2} + \|v_x(t)\|_{H^3} + \|\theta_x(t)\|_{H^2} + \|v_x(t)\|_{L^\infty}\|u_{xxx}(t)\|$$
$$+ \|u_x(t)\|_{L^\infty}\|v_{xxx}(t)\| + \|v_{xx}(t)\|_{L^\infty}\|u_{xx}(t)\|)$$
$$\le C_2(\|u_x(t)\|_{H^2} + \|\theta_x(t)\|_{H^2} + \|v_x(t)\|_{H^3}), \tag{4.4.6}$$

$$\|\theta_{txx}(t)\| \le C_2(\|\theta_x(t)\|_{H^3} + \|u_x(t)\|_{H^2} + \|v_x(t)\|_{H^2} + \|f_{xx}(t)\|)$$
$$\le C_2(\theta_x(t)\|_{H^3} + \|u_x(t)\|_{H^2} + \|v_x(t)\|_{H^2} + \|z_x(t)\|_{H^1}), \tag{4.4.7}$$

$$\|z_{txx}(t)\| \le C_2(\|z_x(t)\|_{H^3} + \|u_x(t)\|_{H^2} + \|f_{xx}(t)\|)$$
$$\le C_2(\|\theta_x(t)\|_{H^1} + \|u_x(t)\|_{H^2} + \|z_x(t)\|_{H^3}) \tag{4.4.8}$$

or

$$\|v_{xxxx}(t)\| \le C_2(\|u_x(t)\|_{H^2} + \|v_x(t)\|_{H^2} + \|\theta_x(t)\|_{H^2} + \|v_{txx}(t)\|), \tag{4.4.9}$$

$$\|\theta_{xxxx}(t)\| \le C_2(\|\theta_x(t)\|_{H^2} + \|u_x(t)\|_{H^2} + \|v_x(t)\|_{H^2}$$
$$+ \|z_x(t)\|_{H^1} + \|\theta_{txx}(t)\|), \tag{4.4.10}$$

$$\|z_{xxxx}(t)\| \le C_2(\|z_x(t)\|_{H^2} + \|u_x(t)\|_{H^2} + \|\theta_x(t)\|_{H^1} + \|z_{txx}(t)\|). \tag{4.4.11}$$

Differentiating (4.1.3) with respect to t, using Theorem 4.1.1, (4.2.37)–(4.2.39), (4.4.6)–(4.4.8) and (4.2.58)–(4.2.60), we have

$$\|v_{tt}(t)\| \leq C_2 \Big(\|v_x(t)\|_{H^1} + \|u_x(t)\| + \|\theta_t(t)\| + \|\theta_{tx}(t)\|$$

$$+ \|v_{tx}(t)\| + \|v_{txx}(t)\| \Big) \tag{4.4.12}$$

$$\leq C_2 \Big(\|u_x(t)\|_{H^2} + \|v_x(t)\|_{H^3} + \|\theta_x(t)\|_{H^2} + \|z_x(t)\| + \|f(t)\| \Big). \tag{4.4.13}$$

Analogously, we get

$$\|\theta_{tt}(t)\| \leq C_2 \Big(\|v_x(t)\|_{H^1} + \|u_x(t)\| + \|\theta_t(t)\|_{H^2} + \|\theta_x(t)\|_{H^2}$$

$$+ \|f_t(t)\| + \|v_{tx}(t)\| \Big) \tag{4.4.14}$$

$$\leq C_2 \Big(\|\theta_x(t)\|_{H^3} + \|u_x(t)\|_{H^2} + \|v_x(t)\|_{H^2} + \|z_x(t)\|_{H^1} + \|f(t)\| \Big), \tag{4.4.15}$$

$$\|z_{tt}(t)\| \leq C_2 \Big(\|z_{txx}(t)\| + \|z_{xx}(t)\| + \|v_x(t)\|_{H^1} + \|\theta_x(t)\|_{H^1}$$

$$+ \|u_x(t)\| + \|f_t(t)\| \Big) \tag{4.4.16}$$

$$\leq C_2 \Big(\|z_x(t)\|_{H^3} + \|v_x(t)\|_{H^1} + \|u_x(t)\|_{H^2} + \|\theta_x(t)\|_{H^1} + \|f(t)\| \Big). \tag{4.4.17}$$

Thus estimate (4.4.2) follows from (4.4.6)–(4.4.8), (4.4.13), (4.4.15) and (4.4.17).

Differentiating (4.1.3) with respect to t twice, multiplying the resulting equation by v_{tt} in $L^2(0,1)$, performing an integration by parts, using Theorem 4.1.1, (4.4.1)–(4.4.2), we obtain

$$\frac{1}{2} \frac{d}{dt} \|v_{tt}(t)\|^2 = - \int_0^1 \left(-\frac{\theta}{u^p} + \mu \frac{v_x}{u} \right)_{tt} v_{ttx} \, dx \tag{4.4.18}$$

$$\leq -\mu \int_0^1 \frac{v_{ttx}^2}{u} \, dx + C_2 \left(\|\theta_{tt}\| + \|v_{tx} v_x\| + \|v_x^3\| + \|\theta_t v_x\| + \|v_{tx}\| + \|v_x^2\| \right) \|v_{ttx}\|$$

$$\leq -C_1^{-1} \|v_{ttx}(t)\|^2 + C_2 \left(\|v_x(t)\|_{H^1}^2 + \|\theta_t(t)\|^2 + \|v_{tx}(t)\|^2 + \|\theta_{tt}(t)\|^2 \right).$$

Thus, by Theorem 4.1.1,

$$\|v_{tt}(t)\|^2 + \int_0^t \|v_{ttx}(\tau)\|^2 \, d\tau \leq C_4 + C_4 \int_0^t \|\theta_{tt}(\tau)\|^2 \, d\tau$$

which, along with (4.4.14) and Lemmas 4.2.1–4.2.3, gives estimate (4.4.3).

Similarly, differentiating (4.1.4) with respect to t twice, multiplying the resulting equation by θ_{tt} in $L^2(0,1)$ and integrating by parts, we arrive at

$$
\frac{1}{2}\frac{d}{dt}\|\theta_{tt}(t)\|^2 = -\int_0^1 \left(\kappa\frac{\theta_x}{u}\right)_{tt}\theta_{ttx}\,dx + \int_0^1 \left(-\frac{\theta}{u^p}+\mu\frac{v_x}{u}\right)v_{ttx}\theta_{tt}\,dx
$$
$$
+ \int_0^1 f_{tt}\theta_{tt}\,dx + \int_0^1 \left(-\frac{\theta}{u^p}+\mu\frac{v_x}{u}\right)_{tt}v_x\theta_{tt}\,dx
$$
$$
+ \int_0^1 \left(-\frac{\theta}{u^p}+\mu\frac{v_x}{u}\right)_t v_{tx}\theta_{tt}\,dx
$$
$$
= A_1 + A_2 + A_3 + A_4 + A_5. \tag{4.4.19}
$$

By virtue of Lemmas 4.2.1–4.2.3 and (4.4.1)–(4.4.2), and using the embedding theorem, we deduce that for any $\varepsilon \in (0,1)$,

$$
\begin{aligned}
A_1 &\le -C_1^{-1}\|\theta_{ttx}(t)\|^2 + C_2(\|v_x(t)\|_{L^\infty}\|\theta_{tx}(t)\| + \|\theta_x(t)\|_{L^\infty}\|v_{tx}(t)\| \\
&\quad + \|v_x(t)\|_{L^\infty}^2\|\theta_x(t)\|)\|\theta_{ttx}(t)\| \\
&\le -(2C_1)^{-1}\|\theta_{ttx}(t)\|^2 + C_2(\|\theta_{tx}(t)\|^2 + \|v_{tx}(t)\|^2 + \|v_x(t)\|_{H^1}^2), \tag{4.4.20}
\end{aligned}
$$
$$
A_2 \le \varepsilon\|v_{ttx}(t)\|^2 + C_2\varepsilon^{-1}\|\theta_{tt}(t)\|^2, \tag{4.4.21}
$$
$$
\begin{aligned}
A_3 &\le \int_0^1 [(|v_x|+|\theta_t|+|z_t|)^2 + |v_{tx}| + |\theta_{tt}| + |z_{tt}|]|\theta_{tt}|\,dx \\
&\le C_1\|\theta_{tt}(t)\|_{L^\infty}\Big[(\|v_x(t)\|_{L^\infty} + \|\theta_t(t)\|_{L^\infty} + \|z_t(t)\|_{L^\infty})(\|v_x(t)\| \\
&\quad + \|\theta_t(t)\| + \|z_t(t)\|)\Big] + C_1(\|v_{tx}(t)\| + \|\theta_{tt}(t)\| + \|z_{tt}(t)\|)\|\theta_{tt}(t)\| \\
&\le C_1(\|\theta_{tt}\| + \|\theta_{ttx}\|)(\|v_x\|_{H^1} + \|\theta_t\|_{H^1} + \|z_t\|_{H^1}) \\
&\quad + C_1(\|v_{tx}(t)\| + \|\theta_{tt}(t)\| + \|z_{tt}(t)\|)\|\theta_{tt}(t)\| \\
&\le \varepsilon(\|\theta_{ttx}(t)\|^2 + \|z_{txx}(t)\|^2) \\
&\quad + C_2\varepsilon^{-1}(\|v_x(t)\|_{H^1}^2 + \|\theta_t(t)\|_{H^1}^2 + \|z_t(t)\|_{H^1}^2 + \|v_{tx}(t)\|^2 + \|\theta_{tt}(t)\|^2), \\
&\tag{4.4.22}
\end{aligned}
$$
$$
\begin{aligned}
A_4 &\le C_2\|v_x(t)\|_{L^\infty}\|\theta_{tt}(t)\|\Big\{\|v_x(t)\|_{L^\infty}(\|\theta_t(t)\| + \|v_x(t)\| + \|v_{tx}(t)\|) \\
&\quad + \|v_{tx}(t)\| + \|\theta_{tt}(t)\| + \|v_{ttx}(t)\| + \|v_x(t)\|_{H^1}\Big\} \\
&\le C_2\|\theta_{tt}(t)\|(\|v_x(t)\|_{H^1} + \|\theta_t(t)\| + \|\theta_{tt}(t)\| + \|v_{tx}(t)\| + \|v_{ttx}(t)\|) \\
&\le \varepsilon\|v_{ttx}(t)\|^2 + C_2\varepsilon^{-1}(\|\theta_{tt}(t)\|^2 + \|v_x(t)\|_{H^1}^2 + \|\theta_t(t)\|^2 + \|v_{tx}(t)\|^2), \tag{4.4.23}
\end{aligned}
$$
$$
\begin{aligned}
A_5 &\le C_2\int_0^1 (|\theta_t| + |v_x| + |v_{tx}| + |v_x^2|)|\theta_{tt}||v_{tx}|\,dx \\
&\le C_2\|v_{tx}(t)\|^{1/2}\|v_{txx}(t)\|^{1/2}(\|v_x(t)\| + \|\theta_t(t)\| + \|v_{tx}(t)\|)\|\theta_{tt}(t)\|
\end{aligned}
$$

which implies

$$\int_0^t A_5 \, d\tau \leq C_2 \sup_{0 \leq \tau \leq t} \|\theta_{tt}(\tau)\| \left(\int_0^t \|v_{txx}(\tau)\|^2 \, d\tau \right)^{1/4} \left(\int_0^t \|v_{tx}(\tau)\|^2 \, d\tau \right)^{1/4}$$

$$\times \left\{ \int_0^t (\|v_x\|^2 + \|\theta_t\|^2 + \|v_{tx}\|^2)(\tau) \, d\tau \right\}^{1/2}$$

$$\leq \varepsilon \left(\sup_{0 \leq \tau \leq t} \|\theta_{tt}(\tau)\|^2 + \int_0^t \|v_{txx}(\tau)\|^2 d\tau \right) + C_2 \varepsilon^{-3}. \tag{4.4.24}$$

Thus we infer from (4.4.19)–(4.4.24) that for $\varepsilon \in (0,1)$ small enough,

$$\|\theta_{tt}(t)\|^2 + \int_0^t \|\theta_{ttx}(\tau)\|^2 \, d\tau$$

$$\leq C_4 \varepsilon^{-3} + C_2 \varepsilon^{-1} \int_0^t \|\theta_{tt}(\tau)\|^2 \, d\tau \tag{4.4.25}$$

$$+ C_1 \varepsilon \left\{ \sup_{0 \leq \tau \leq t} \|\theta_{tt}(\tau)\|^2 + \int_0^t (\|v_{txx}\|^2 + \|z_{txx}\|^2 + \|v_{ttx}\|^2)(\tau) \, d\tau \right\}.$$

Thus taking supremum in t on the left-hand side of (4.4.25), picking $\varepsilon \in (0,1)$ small enough, and using (4.4.14), we can derive estimate (4.4.4).

Differentiating (4.1.5) with respect to t twice, multiplying the resulting equation by z_{tt} in $L^2(0,1)$ and integrating by parts, we obtain

$$\frac{1}{2} \frac{d}{dt} \|z_{tt}\|^2 = -\int_0^1 \left(\frac{d}{u^2} z_x \right)_{tt} z_{ttx} \, dx - \int_0^1 f_{tt} z_{tt} \, dx$$

$$\leq -\int_0^1 d \frac{z_{ttx}^2}{u^2} \, dx + C_2 \Big[\big(\|z_{tx}(t)\|$$

$$+ \|v_{tx}(t)\| + \|z_x(t)\| \big) \|z_{ttx}(t)\| + \|f_{tt}(t)\| \|z_{tt}(t)\| \Big]$$

$$\leq -(2C_1)^{-1} \|z_{ttx}(t)\|^2$$

$$+ C_2 \Big(\|z_{tt}(t)\|^2 + \|f_{tt}(t)\|^2 + \|z_{tx}(t)\|^2 + \|v_{tx}(t)\|^2 + \|z_x(t)\|^2 \Big)$$

$$\leq -(2C_1)^{-1} \|z_{ttx}(t)\|^2 + C_2 \Big(\|z_{tt}(t)\|^2 + \|\theta_{tt}(t)\|^2 + \|v_{tx}(t)\|^2$$

$$+ \|v_x(t)\|_{H^1}^2 + \|\theta_t(t)\|_{H^1}^2 + \|z_t(t)\|_{H^1}^2 + \|z_x(t)\|^2 \Big). \tag{4.4.26}$$

Thus, by Lemmas 4.2.1–4.2.3 and (4.4.2),

$$\|z_{tt}(t)\|^2 + \int_0^t \|z_{ttx}(\tau)\|^2 \, d\tau \leq C_4 + C_2 \int_0^t (\|z_{tt}\|^2 + \|\theta_{tt}\|^2)(\tau) \, d\tau \tag{4.4.27}$$

which, together with (4.4.14) and (4.4.16), gives (4.4.5). The proof is complete. $\qquad\square$

Lemma 4.4.2. *Under assumptions of Theorem 4.1.2, for any $(u_0, v_0, \theta_0, z_0) \in H_+^4$, the following estimates hold for any $t > 0$,*

$$\|v_{tx}(t)\|^2 + \int_0^t \|v_{txx}(\tau)\|^2 \, d\tau \leq C_4 + C_1 \varepsilon^2 \int_0^t (\|v_{ttx}\|^2 + \|\theta_{txx}\|^2)(\tau) \, d\tau, \quad (4.4.28)$$

$$\|\theta_{tx}(t)\|^2 + \int_0^t \|\theta_{txx}(\tau)\|^2 \, d\tau \leq C_4 + C_2 \varepsilon^2 \int_0^t \|v_{txx}(\tau)\|^2 \, d\tau, \quad (4.4.29)$$

$$\|z_{tx}(t)\|^2 + \int_0^t \|z_{txx}(\tau)\|^2 \, d\tau \leq C_4 \quad (4.4.30)$$

for $\varepsilon \in (0,1)$ small enough.

Proof. Differentiating (4.1.3) with respect to x and t, multiplying the resulting equation by v_{tx} in $L^2(0,1)$, and integrating by parts, we have

$$\frac{1}{2}\frac{d}{dt}\|v_{tx}(t)\|^2 = B_0(t) + B_1(t) \quad (4.4.31)$$

with

$$B_0(t) = \left(-\frac{\theta}{u^p} + \mu\frac{v_x}{u}\right)_{tx} v_{tx}\Big|_{x=0}^{x=1}, \qquad B_1(t) = -\int_0^1 \left(-\frac{\theta}{u^p} + \mu\frac{v_x}{u}\right)_{tx} v_{txx} \, dx.$$

We employ Lemmas 4.2.1–4.2.3 and Lemma 4.4.1, the interpolation inequality and Poincaré's inequality to get

$$B_0(t) \leq C_1 \Big[(\|v_x(t)\|_{L^\infty} + \|\theta_t(t)\|_{L^\infty})(\|u_x(t)\|_{L^\infty} + \|\theta_x(t)\|_{L^\infty})$$
$$+ \|v_{xx}(t)\|_{L^\infty} + \|\theta_{tx}(t)\|_{L^\infty} + \|v_{txx}(t)\|_{L^\infty} + \|u_x(t)\|_{L^\infty}\|v_{tx}(t)\|_{L^\infty}$$
$$+ \|v_x(t)\|_{L^\infty}\|v_{xx}(t)\|_{L^\infty} + \|v_x^2(t)\|_{L^\infty}\Big]\|v_{tx}(t)\|_{L^\infty}$$
$$\leq C_2(B_{01} + B_{02})\|v_{tx}(t)\|^{1/2}\|v_{txx}(t)\|^{1/2} \quad (4.4.32)$$

where

$$B_{01} = \|v_x(t)\|_{H^2} + \|\theta_t(t)\| + \|\theta_{tx}(t)\|$$

and

$$B_{02} = \|\theta_{tx}(t)\|^{1/2}\|\theta_{txx}(t)\|^{1/2} + \|v_{txx}(t)\|^{1/2}\|v_{txxx}(t)\|^{1/2} + \|v_{txx}(t)\|$$
$$+ \|v_{tx}(t)\|^{1/2}\|v_{txx}(t)\|^{1/2}.$$

Applying Young's inequality several times, we have that for $\varepsilon \in (0,1)$,

$$C_2 B_{01}\|v_{tx}(t)\|^{1/2}\|v_{txx}(t)\|^{1/2}$$
$$\leq \frac{\varepsilon^2}{2}\|v_{txx}(t)\|^2 + C_2 \varepsilon^{-1}\Big(\|v_{tx}(t)\|^2 + \|v_x(t)\|_{H^2}^2 + \|\theta_t(t)\|_{H^1}^2\Big) \quad (4.4.33)$$

and

$$C_2 B_{02} \|v_{tx}(t)\|^{1/2} \|v_{txx}(t)\|^{1/2} \leq \frac{\varepsilon^2}{2} \|v_{txx}(t)\|^2 + \varepsilon^2 \left(\|\theta_{txx}(t)\|^2 + \|v_{txxx}(t)\|^2 \right)$$
$$+ C_2 \varepsilon^{-6} (\|\theta_{tx}(t)\|^2 + \|v_{tx}(t)\|^2). \qquad (4.4.34)$$

Thus we infer from (4.4.32)–(4.4.34) and Lemmas 4.2.1–4.2.3, Lemma 4.4.1 that

$$B_0 \leq \varepsilon^2 (\|v_{txx}(t)\|^2 + \|v_{txxx}(t)\|^2 + \|\theta_{txx}(t)\|^2)$$
$$+ C_2 \varepsilon^{-6} (\|v_x(t)\|_{H^2}^2 + \|\theta_x(t)\|^2 + \|\theta_{tx}(t)\|^2 + \|v_{tx}(t)\|^2) \qquad (4.4.35)$$

which, together with Theorem 4.1.1, further leads to

$$\int_0^t B_0 \, d\tau \leq \varepsilon^2 \int_0^t (\|v_{txx}\|^2 + \|v_{txxx}\|^2 + \|\theta_{txx}\|^2)(\tau) \, d\tau + C_2 \varepsilon^{-6}, \quad \forall t > 0.$$
$$(4.4.36)$$

Analogously, by Lemma 4.4.1 and Theorem 4.1.1 and the embedding theorem, we get that for any $\varepsilon \in (0,1)$,

$$B_1 \leq -\mu \int_0^1 \frac{v_{txx}^2}{u} \, dx + C_1 \Big\{ (\|v_x\| + \|\theta_t\|)(\|u_x\|_{L^\infty} + \|\theta_x\|_{L^\infty}) \qquad (4.4.37)$$
$$+ \|v_{xx}\| + \|\theta_{tx}\| + \|u_x\|_{L^\infty} \|v_{tx}\| + \|v_x\|_{L^\infty} \|v_{xx}\| + \|v_x\|_{L^\infty}^2 \|u_x\| \Big\} \|v_{txx}\|$$
$$\leq -(2C_1)^{-1} \|v_{txx}(t)\|^2 + C_2(\|v_x(t)\|_{H^1}^2 + \|\theta_t(t)\|_{H^1}^2 + \|v_{tx}(t)\|^2 + \|u_x(t)\|^2)$$

which, combined with (4.4.31), (4.4.36) and Theorem 4.1.1, gives that for $\varepsilon \in (0,1)$ small enough,

$$\|v_{tx}(t)\|^2 + \int_0^t \|v_{txx}(\tau)\|^2 \, d\tau \leq C_3 \varepsilon^{-6} + C_1 \varepsilon^2 \int_0^t (\|\theta_{txx}\|^2 + \|v_{txxx}\|^2)(\tau) \, d\tau.$$
$$(4.4.38)$$

On the other hand, differentiating (4.1.3) with respect to x and t, and using Theorem 4.1.1 and Lemma 4.4.1, we derive

$$\|v_{txxx}(t)\| \leq C_1 \|v_{ttx}(t)\| + C_2(\|v_x(t)\|_{H^2} + \|\theta_x(t)\|_{H^1} + \|u_x(t)\|_{H^1} + \|\theta_t(t)\|_{H^2}).$$
$$(4.4.39)$$

Thus inserting (4.4.39) into (4.4.38) implies (4.4.28).

Similarly, we derive from (4.1.4) that

$$\frac{1}{2} \frac{d}{dt} \|\theta_{tx}(t)\|^2 = D_1(t) + D_2(t) + D_3(t) \qquad (4.4.40)$$

with

$$D_1(t) = -\int_0^1 \left(\frac{\kappa}{u}\theta_x\right)_{tx}\theta_{txx}\,dx,$$

$$D_2(t) = \int_0^1 \left(\left(-\frac{\theta}{u^p} + \mu\frac{v_x}{u}\right)v_x\right)_{tx}\theta_{tx}\,dx,$$

$$D_3(t) = \int_0^1 f_{tx}\theta_{tx}\,dx.$$

By virtue of Lemmas 4.2.1–4.2.3, Lemma 4.4.1 and (4.4.28), and using the embedding theorem, we deduce that for any $\varepsilon \in (0,1)$,

$$D_1 \leq -(2C_1)^{-1}\|\theta_{txx}(t)\|^2 + C_2\left(\|v_x(t)\|_{H^1}^2 + \|\theta_x(t)\|_{H^1}^2 + \|\theta_t(t)\|_{H^1}^2\right), \quad (4.4.41)$$

$$D_2 \leq \varepsilon^2\|v_{txx}(t)\|^2$$
$$+ C_2\varepsilon^{-2}\left(\|v_x(t)\|_{H^2}^2 + \|\theta_t(t)\|_{H^1}^2 + \|v_{tx}(t)\|^2 + \|u_x(t)\|_{H^1}^2\right), \quad (4.4.42)$$

$$D_3 \leq C_2(\|f_{tx}(t)\|^2 + \|\theta_{tx}(t)\|^2)$$

which, along with (4.4.40)–(4.4.42), implies (4.4.29).

Differentiating (4.1.5) with respect to x and t, multiplying the resulting equation by z_{tx}, and integrating by parts, we arrive at

$$\frac{1}{2}\frac{d}{dt}\|z_{tx}(t)\|^2 = -\int_0^1 \left(\frac{d}{u^2}z_x\right)_{tx}z_{txx}\,dx - \int_0^1 f_{tx}z_{tx}\,dx \quad (4.4.43)$$

$$\leq -d\int_0^1 \frac{z_{txx}^2}{u^2}\,dx + C_1\left(\|z_{tx}(t)\| + \|z_{xx}(t)\| + \|v_x(t)\|_{H^1}\right)\|z_{txx}(t)\| + \|z_{tx}(t)\|\|f_{tx}(t)\|$$

$$\leq -(2C_1)^{-1}\|z_{txx}(t)\|^2 + C_2\left(\|v_x(t)\|_{H^1}^2 + \|z_{xx}(t)\|^2 + \|f_{tx}(t)\|_{H^1}^2 + \|z_{tx}(t)\|^2\right)$$

which, together with Lemmas 4.2.1–4.2.3 and Lemma 4.4.1, gives (4.4.30). The proof is complete. □

Lemma 4.4.3. *Under assumptions of Theorem 4.1.2, for any $(u_0, v_0, \theta_0, z_0) \in H_+^4$, we have for any $t > 0$,*

$$\|v_{tt}(t)\|^2 + \|v_{tx}(t)\|^2 + \|\theta_{tt}(t)\|^2 + \|\theta_{tx}(t)\|^2 + \|z_{tt}(t)\|^2 + \|z_{tx}(t)\|^2 + \int_0^t \left(\|v_{ttx}\|^2\right.$$

$$+ \|v_{txx}\|^2 + \|\theta_{ttx}\|^2 + \|\theta_{txx}\|^2 + \|z_{ttx}\|^2 + \|z_{txx}\|^2\right)(\tau)\,d\tau \leq C_4, \quad (4.4.44)$$

$$\|u_{xxx}(t)\|_{H^1}^2 + \|v_{xxx}(t)\|_{H^1}^2 + \|\theta_{xxx}(t)\|_{H^1}^2 + \|z_{xxx}(t)\|_{H^1}^2 + \|v_{txx}(t)\|^2$$

$$+ \|\theta_{txx}(t)\|^2 + \|z_{txx}(t)\|^2 + \int_0^t \left(\|v_{tt}\|^2 + \|\theta_{tt}\|^2 + \|z_{tt}\|^2 + \|v_{txx}\|_{H^1}^2\right.$$

$$+ \|\theta_{txx}\|_{H^1}^2 + \|z_{txx}\|_{H^1}^2\right)(\tau)\,d\tau \leq C_4, \quad (4.4.45)$$

$$\int_0^t \left(\|u_{xxx}\|_{H^1}^2 + \|v_{xxxx}\|_{H^1}^2 + \|\theta_{xxxx}\|_{H^1}^2 + \|z_{xxxx}\|_{H^1}^2\right)(\tau)\,d\tau \leq C_4. \quad (4.4.46)$$

Proof. Adding up (4.4.28), (4.4.29) and (4.4.30), choosing ε small enough, we conclude

$$\|v_{tx}(t)\|^2 + \|\theta_{tx}(t)\|^2 + \|z_{tx}(t)\|^2 + \int_0^t (\|v_{txx}\|^2 + \|\theta_{txx}\|^2 + \|z_{txx}\|^2)(\tau)\,d\tau$$

$$\leq C_4 + C_2\varepsilon^2 \int_0^t \|v_{ttx}(\tau)\|^2\,d\tau. \tag{4.4.47}$$

Now multiplying (4.4.3), (4.4.4) and (4.4.5) by ε, ε^2 and ε, respectively; then adding the resultant to (4.4.47), and picking ε small enough, we obtain (4.4.44).

Differentiating (4.2.56) with respect to x, we get

$$\mu\frac{\partial}{\partial t}\left(\frac{u_{xxx}}{u}\right) - \mathcal{P}_u u_{xxx} = E_1(x,t) \tag{4.4.48}$$

where

$$E_1(x,t) = v_{txx} + E_x(x,t) + \mathcal{P}_{ux} u_{xx} + \mu\left(\frac{u_{xx}u_x}{u^2}\right)_t.$$

Obviously, we can infer from Lemmas 4.2.1–4.2.3 and Lemmas 4.4.1–4.4.2 that

$$\|E_1(t)\| \leq C_2(\|u_x(t)\|_{H^1} + \|v_x(t)\|_{H^2} + \|\theta_x(t)\|_{H^2} + \|v_{txx}(t)\|) \tag{4.4.49}$$

leading to

$$\int_0^t \|E_1(\tau)\|^2\,d\tau \leq C_4, \quad \forall t > 0. \tag{4.4.50}$$

Multiplying (4.4.48) by $\frac{u_{xxx}}{u}$ in $L^2(0,1)$, we obtain

$$\frac{d}{dt}\left\|\frac{u_{xxx}}{u}\right\|^2 + C_1^{-1}\left\|\frac{u_{xxx}}{u}\right\|^2 \leq C_1\|E_1(t)\|^2 \tag{4.4.51}$$

which, combined with (4.4.50), gives

$$\|u_{xxx}(t)\|^2 + \int_0^t \|u_{xxx}(\tau)\|^2\,d\tau \leq C_4, \quad \forall t > 0. \tag{4.4.52}$$

By (4.2.61)–(4.2.63), (4.4.9)–(4.4.11), (4.4.44), (4.4.52), Lemmas 4.2.1–4.2.3 and Lemmas 4.4.1–4.4.2, and using the embedding theorem, we obtain for any $t > 0$,

$$\|v_{xxx}(t)\|^2 + \|\theta_{xxx}(t)\|^2 + \|z_{xxx}(t)\|^2 + \|v_{xx}(t)\|_{L^\infty}^2 + \|\theta_{xx}(t)\|_{L^\infty}^2 + \|z_{xx}(t)\|_{L^\infty}^2$$

$$+ \int_0^t \Big(\|v_{xxx}\|_{H^1}^2 + \|\theta_{xxx}\|_{H^1}^2 + \|z_{xxx}\|_{H^1}^2 + \|v_{xx}\|_{W^{1,\infty}}^2$$

$$+ \|\theta_{xx}\|_{W^{1,\infty}}^2 + \|z_{xx}\|_{W^{1,\infty}}^2\Big)(\tau)\,d\tau \leq C_4. \tag{4.4.53}$$

Differentiating (4.1.3)–(4.1.5) with respect to t, using (4.4.44), (4.4.52)–(4.4.53), Lemmas 4.4.1–4.4.2 and Lemmas 4.2.1–4.2.2, we conclude

$$
\begin{aligned}
\|v_{txx}(t)\| &\leq C_1\|v_{tt}(t)\| + C_2(\|v_x(t)\|_{H^1} + \|u_x(t)\| + \|\theta_t(t)\| \\
&\quad + \|\theta_{tx}(t)\| + \|v_{tx}(t)\|) \leq C_4,
\end{aligned} \tag{4.4.54}
$$

$$
\begin{aligned}
\|\theta_{txx}(t)\| &\leq C_1\|\theta_{tt}(t)\| + C_2(\|v_x(t)\|_{H^1} + \|u_x(t)\| + \|\theta_t(t)\|_{H^1} + \|\theta_x(t)\|_{H^2} \\
&\quad + \|f_t(t)\| + \|v_{tx}(t)\|) \leq C_4,
\end{aligned} \tag{4.4.55}
$$

$$
\begin{aligned}
\|z_{txx}(t)\| &\leq C_1\|z_{tt}(t)\| + C_2(\|z_{xx}(t)\| + \|v_x(t)\|_{H^1} + \|\theta_x(t)\|_{H^1} \\
&\quad + \|u_x(t)\| + \|f_t(t)\|) \leq C_4
\end{aligned} \tag{4.4.56}
$$

which, combined with (4.4.9)–(4.4.11), implies

$$
\begin{aligned}
\|v_{xxxx}(t)\|^2 &+ \|\theta_{xxxx}(t)\|^2 + \|z_{xxxx}(t)\|^2 + \int_0^t (\|v_{txx}\|^2 + \|z_{txx}\|^2 + \|\theta_{txx}\|^2 \\
&+ \|v_{xxxx}\|^2 + \|\theta_{xxxx}\|^2 + \|z_{xxxx}\|^2)(\tau)\, d\tau \leq C_4, \quad \forall t > 0.
\end{aligned} \tag{4.4.57}
$$

Therefore it follows from (4.4.53), (4.4.57) and the embedding theorem that

$$
\begin{aligned}
\|v_{xxx}(t)\|_{L^\infty}^2 &+ \|\theta_{xxx}(t)\|_{L^\infty}^2 + \|z_{xxx}(t)\|_{L^\infty}^2 + \int_0^t (\|v_{xxx}\|_{L^\infty}^2 \\
&+ \|\theta_{xxx}\|_{L^\infty}^2 + \|z_{xxx}\|_{L^\infty}^2)(\tau)\, d\tau \leq C_4, \quad \forall t > 0.
\end{aligned} \tag{4.4.58}
$$

Now differentiating (4.4.48) with respect to x, we find

$$
\mu \frac{\partial}{\partial t}\left(\frac{u_{xxxx}}{u}\right) - \mathcal{P}_u u_{xxxx} = E_2(x,t) \tag{4.4.59}
$$

where

$$
E_2(x,t) = E_{1x}(x,t) + \mathcal{P}_{ux} u_{xxx} + \mu\left(\frac{u_{xxx}u_x}{u^2}\right)_t.
$$

Using the embedding theorem, Lemmas 4.4.1–4.4.2, (4.4.53) and (4.4.54)–(4.4.58), we derive that

$$
\|E_{xx}(t)\| \leq C_4(\|\theta_x(t)\|_{H^3} + \|u_x(t)\|_{H^2} + \|v_x(t)\|_{H^3}),
$$

$$
\begin{aligned}
\|E_{1x}(t)\| &\leq C_1\left(\|E_{xx}(t)\| + \|v_{txxx}(t)\| + \|(p_{ux}u_{xx})_x(t)\| + \left\|\left(\frac{u_{xx}u_x}{u^2}\right)_{tx}(t)\right\|\right) \\
&\leq C_1\|v_{txxx}(t)\| + C_4(\|\theta_x(t)\|_{H^3} + \|u_x(t)\|_{H^2} + \|v_x(t)\|_{H^3}),
\end{aligned}
$$

whence

$$
\|E_2(t)\| \leq C_1\|v_{txxx}(t)\| + C_4(\|\theta_x(t)\|_{H^3} + \|u_x(t)\|_{H^2} + \|v_x(t)\|_{H^3}). \tag{4.4.60}
$$

We infer from (4.4.12)–(4.4.17) that

$$
\int_0^t (\|v_{tt}\|^2 + \|\theta_{tt}\|^2 + \|z_{tt}\|^2)(\tau)\, d\tau \leq C_4, \quad \forall t > 0 \tag{4.4.61}
$$

which, together with (4.4.39) and (4.4.44), gives

$$\int_0^t \|v_{txxx}(\tau)\|^2\, d\tau \le C_4, \quad \forall\, t > 0. \tag{4.4.62}$$

Thus it follows from (4.4.53), (4.4.57), (4.4.60), (4.4.62), Lemmas 4.2.1–4.2.2 and Lemmas 4.4.1–4.4.2 that

$$\int_0^t \|E_2(\tau)\|^2\, d\tau \le C_4, \quad \forall\, t > 0. \tag{4.4.63}$$

Multiplying (4.4.59) by $\frac{u_{xxxx}}{u}$ in $L^2(0,1)$, we get

$$\frac{d}{dt}\left\|\frac{u_{xxxx}}{u}\right\|^2 + C_1^{-1}\left\|\frac{u_{xxxx}}{u}\right\|^2 \le C_1\|E_2\|^2, \tag{4.4.64}$$

whence by (4.4.63),

$$\|u_{xxxx}(t)\|^2 + \int_0^t \|u_{xxxx}(\tau)\|^2\, d\tau \le C_4, \quad \forall\, t > 0. \tag{4.4.65}$$

Differentiating (4.1.4) and (4.1.5) with respect to x and t, respectively, and using Lemmas 4.2.1–4.2.3 and Lemmas 4.4.1–4.4.2, we derive

$$\|\theta_{txxx}(t)\| \le C_1\|\theta_{ttx}(t)\| + C_2(\|v_x(t)\|_{H^3} + \|u_x(t)\|_{H^2} $$
$$+ \|\theta_x(t)\|_{H^3} + \|f_{tx}(t)\|), \tag{4.4.66}$$
$$\|z_{txxx}(t)\| \le C_1\|z_{ttx}(t)\| + C_2(\|v_x(t)\|_{H^3} + \|u_x(t)\|_{H^3} $$
$$+ \|z_x(t)\|_{H^2} + \|f_{tx}(t)\|). \tag{4.4.67}$$

Thus,

$$\int_0^t (\|\theta_{txxx}\|^2 + \|z_{txxx}\|^2)(\tau)\, d\tau \le C_4, \quad \forall\, t > 0. \tag{4.4.68}$$

Differentiating (4.1.3) with respect to x three times, using Lemmas 4.4.1–4.4.2 and Lemmas 4.2.1–4.2.3, applying Poincaré's inequality, we deduce

$$\|v_{xxxxx}(t)\| \le C_1\|v_{txxx}(t)\| + C_2(\|u_x(t)\|_{H^3} + \|v_x(t)\|_{H^3} + \|\theta_x(t)\|_{H^3}). \tag{4.4.69}$$

Thus we derive from (4.4.52)–(4.4.53), (4.4.57)–(4.4.58), (4.4.62) and (4.4.65) that

$$\int_0^t \|v_{xxxxx}(\tau)\|^2\, d\tau \le C_4, \quad \forall\, t > 0. \tag{4.4.70}$$

Similarly, we can deduce from (4.1.4)–(4.1.5), (4.4.65), (4.4.68) and (4.4.57) that

$$\int_0^t (\|\theta_{xxxxx}\|^2 + \|z_{xxxxx}\|^2)(\tau)\, d\tau \le C_4, \quad \forall\, t > 0. \tag{4.4.71}$$

Hence, (4.4.45)–(4.4.46) follow from (4.4.52)–(4.4.71). The proof is complete. $\square$

Lemma 4.4.4. *Under assumptions of Theorem 4.1.2, for any $(u_0, v_0, \theta_0, z_0) \in H_+^4$, there exists a positive constant $\gamma_4^{(1)} = \gamma_4^{(1)}(C_4) \leq \gamma_2(C_2)$ such that for any fixed $\gamma \in (0, \gamma_4^{(1)}]$, the following estimates hold for any $t > 0$ and $\varepsilon \in (0, 1)$ small enough:*

$$e^{\gamma t}\|v_{tt}(t)\|^2 + \int_0^t e^{\gamma\tau}\|v_{ttx}(\tau)\|^2 d\tau \leq C_4 + C_2 \int_0^t e^{\gamma\tau}\|\theta_{txx}(\tau)\|^2 d\tau, \tag{4.4.72}$$

$$e^{\gamma t}\|\theta_{tt}(t)\|^2 + \int_0^t e^{\gamma\tau}\|\theta_{ttx}(\tau)\|^2 d\tau \leq C_4\varepsilon^{-3} + C_2\varepsilon^{-1} \int_0^t e^{\gamma\tau}(\|\theta_{txx}\|^2 + \|z_{txx}\|^2)(\tau) d\tau$$

$$+ \varepsilon \int_0^t e^{\gamma\tau}(\|v_{txx}\|^2 + \|v_{ttx}\|^2)(\tau) d\tau, \tag{4.4.73}$$

$$e^{\gamma t}\|z_{tt}(t)\|^2 + \int_0^t e^{\gamma\tau}\|z_{ttx}(\tau)\|^2 d\tau \leq C_4 + C_2 \int_0^t e^{\gamma\tau}(\|\theta_{txx}\|^2 + \|z_{txx}\|^2)(\tau) d\tau. \tag{4.4.74}$$

Proof. Multiplying (4.4.18) by $e^{\gamma t}$ and using (4.4.14) and Theorem 4.1.1, we have

$$\frac{1}{2}e^{\gamma t}\|v_{tt}(t)\|^2 \leq C_4 - (C_1^{-1} - C_1\gamma) \int_0^t e^{\gamma\tau}\|v_{ttx}(\tau)\|^2 d\tau + C_2 \int_0^t e^{\gamma\tau}\|\theta_{tt}(\tau)\|^2 d\tau$$

$$\leq C_4 - (C_1^{-1} - C_1\gamma) \int_0^t e^{\gamma\tau}\|v_{ttx}(\tau)\|^2 d\tau + C_4 \int_0^t e^{\gamma\tau}\|\theta_{txx}(\tau)\|^2 d\tau$$

which gives (4.4.72) if we take $\gamma > 0$ so small that $0 < \gamma \leq \min\left[\frac{1}{4C_1^2}, \gamma_2(C_2)\right]$. Similarly, multiplying (4.4.19) by $e^{\gamma t}$ and using (4.4.14), (4.4.16) and Theorem 4.1.1, we have

$$\frac{1}{2}e^{\gamma t}\|\sqrt{e_\theta}\theta_{tt}\|^2$$

$$\leq C_4 + \frac{\gamma}{2} \int_0^t e^{\gamma\tau}\|\sqrt{e_\theta}\theta_{tt}\|^2(\tau) d\tau + \int_0^t e^{\gamma\tau}(A_1 + A_2 + A_3 + A_4 + A_5 + A_6 + A_7)$$

$$\leq C_4\varepsilon^{-3} - (C_1^{-1} - 2\varepsilon) \int_0^t e^{\gamma\tau}\|\theta_{ttx}\|^2(\tau) d\tau$$

$$+ C_2\varepsilon^{-1} \int_0^t e^{\gamma\tau}(\|\theta_{txx}\|^2 + \|z_{txx}\|^2)(\tau) d\tau + \varepsilon \int_0^t e^{\gamma\tau}\|v_{txx}\|^2(\tau) d\tau$$

$$+ C_2 e^{\gamma t/2} \sup_{0\leq\tau\leq t}\|\theta_{tt}(\tau)\| \left(\int_0^t e^{\gamma\tau}\|v_{txx}\|^2(\tau) d\tau\right)^{1/4}$$

$$\times \left[\int_0^t e^{\gamma\tau}(\|v_{tx}\|^2 + \|v_x\|^2 + \|\theta_t\|^2)(\tau) d\tau\right]^{1/2} \left(\int_0^t e^{\gamma\tau}\|v_{tx}\|^2(\tau) d\tau\right)^{1/4}$$

$$\leq C_4\varepsilon^{-3} - (C_1^{-1} - 2\varepsilon) \int_0^t e^{\gamma\tau}\|\theta_{ttx}\|^2 d\tau + C_2\varepsilon^{-1} \int_0^t e^{\gamma\tau}(\|\theta_{txx}\|^2$$

$$+ \|z_{txx}\|^2)(\tau) d\tau + \varepsilon \int_0^t e^{\gamma\tau}(\|v_{txx}\|^2 + \|v_{ttx}\|^2)(\tau) d\tau + \varepsilon e^{\gamma t} \sup_{0\leq\tau\leq t}\|\theta_{tt}(\tau)\|^2$$

which, by taking supremum on the right-hand side and choosing $\varepsilon \in (0,1)$ small enough, implies (4.4.73).

Multiplying (4.4.26) by $e^{\gamma t}$ and using (4.4.14), (4.4.16) and Theorem 4.1.1, we have

$$\frac{1}{2}e^{\gamma t}\|z_{tt}(t)\|^2 + C_1^{-1}\int_0^t e^{\gamma\tau}\|z_{ttx}(\tau)\|^2 d\tau \leq C_4 + C_2 \int_0^t e^{\gamma\tau}(\|z_{tt}\|^2 + \|\theta_{tt}\|^2)(\tau)d\tau$$

$$\leq C_4 + C_2 \int_0^t e^{\gamma\tau}(\|z_{txx}\|^2 + \|\theta_{txx}\|^2)(\tau)d\tau$$

which implies (4.4.74). The proof is complete. $\qquad\square$

Lemma 4.4.5. *Under assumptions of Theorem 4.1.2, for any $(u_0, v_0, \theta_0, z_0) \in H_+^4$, there exists a positive constant $\gamma_4^{(2)} \leq \gamma_4^{(1)}$ such that for any fixed $\gamma \in (0, \gamma_4^{(2)}]$, the following estimates hold for any $t > 0$ and $\varepsilon \in (0,1)$ small enough:*

$$e^{\gamma t}\|v_{tx}(t)\|^2 + \int_0^t e^{\gamma\tau}\|v_{txx}(\tau)\|^2 d\tau \leq C_4 + C_2\varepsilon^2 \int_0^t e^{\gamma\tau}(\|v_{ttx}\|^2 + \|\theta_{txx}\|^2)(\tau)d\tau,$$
$$(4.4.75)$$

$$e^{\gamma t}\|\theta_{tx}(t)\|^2 + \int_0^t e^{\gamma\tau}\|\theta_{txx}(\tau)\|^2 d\tau \leq C_4 + \varepsilon^2 \int_0^t e^{\gamma\tau}(\|\theta_{ttx}\|^2 + \|v_{txx}\|^2)(\tau)d\tau,$$
$$(4.4.76)$$

$$e^{\gamma t}\|z_{tx}(t)\|^2 + \int_0^t e^{\gamma\tau}\|z_{txx}(\tau)\|^2 d\tau \leq C_4,$$
$$(4.4.77)$$

$$e^{\gamma t}(\|v_{tx}(t)\|^2 + \|\theta_{tx}(t)\|^2 + \|z_{tx}(t)\|^2) + \int_0^t e^{\gamma\tau}(\|v_{txx}\|^2 + \|\theta_{txx}\|^2 + \|z_{txx}\|^2)(\tau)d\tau$$

$$\leq C_4 + C_2\varepsilon^2 \int_0^t e^{\gamma\tau}(\|v_{ttx}\|^2 + \|\theta_{ttx}\|^2)(\tau)d\tau.$$
$$(4.4.78)$$

Proof. Multiplying (4.4.31) by $e^{\gamma t}$ and using (4.4.35), (4.4.37) and Theorem 4.1.1, we infer that for any $\varepsilon \in (0,1)$ small enough,

$$e^{\gamma t}\|v_{tx}(t)\|^2 \leq C_4 - [(2C_1)^{-1} - \varepsilon^2]\int_0^t e^{\gamma\tau}\|v_{txx}(\tau)\|^2 d\tau$$

$$+ \varepsilon^2 \int_0^t e^{\gamma\tau}(\|v_{txxx}\|^2 + \|\theta_{txx}\|^2)(\tau)d\tau$$

which, together with (4.4.39), gives (4.4.75) if we take $\gamma > 0$ and $\varepsilon \in (0,1)$ so small enough that $0 < \varepsilon < \min[1, 1/(8C_1)]$ and $0 < \gamma \leq \min[\gamma_4^{(1)}, 1/(8C_1^2)] \equiv \gamma_4^{(2)}$.

In the same manner, we easily derive (4.4.76) from (4.4.40)–(4.4.42). And we deduce (4.4.77) from (4.4.43) and Theorem 4.1.1. Adding (4.4.75), (4.4.76) to (4.4.77) and picking $\varepsilon \in (0,1)$ small enough give (4.4.78). The proof is complete. $\qquad\square$

Lemma 4.4.6. *Under assumptions of Theorem 4.1.2, for any $(u_0, v_0, \theta_0, z_0) \in H^4_+$, there is a positive constant $\gamma_4 \leq \gamma_4^{(2)}$ such that for any fixed $\gamma \in (0, \gamma_4]$, the following estimates hold for any $t > 0$:*

$$e^{\gamma t}(\|v_{tt}(t)\|^2 + \|\theta_{tt}(t)\|^2 + \|z_{tt}(t)\|^2 + \|v_{tx}(t)\|^2 + \|\theta_{tx}(t)\|^2 + \|z_{tx}(t)\|^2) \qquad (4.4.79)$$

$$+ \int_0^t e^{\gamma \tau}(\|v_{ttx}\|^2 + \|\theta_{ttx}\|^2 + \|z_{ttx}\|^2 + \|v_{txx}\|^2 + \|\theta_{txx}\|^2 + \|z_{txx}\|^2)(\tau)d\tau \leq C_4,$$

$$e^{\gamma t}\|u_{xxx}(t)\|^2_{H^1} + \int_0^t e^{\gamma \tau}\|u_{xxx}(\tau)\|^2_{H^1}d\tau \leq C_4, \qquad (4.4.80)$$

$$e^{\gamma t}(\|v_{xxx}(t)\|^2_{H^1} + \|\theta_{xxx}(t)\|^2_{H^1} + \|z_{xxx}(t)\|^2_{H^1} + \|v_{txx}(t)\|^2 + \|\theta_{txx}(t)\|^2 + \|z_{txx}(t)\|^2)$$

$$+ \int_0^t e^{\gamma \tau}\Big(\|v_{xxxx}\|^2_{H^1} + \|\theta_{xxxx}\|^2_{H^1} + \|z_{xxxx}\|^2_{H^1} + \|v_{txx}\|^2_{H^1}$$

$$+ \|\theta_{txx}\|^2_{H^1} + \|z_{txx}\|^2_{H^1} + \|v_{tt}\|^2 + \|\theta_{tt}\|^2 + \|z_{tt}\|^2\Big)(\tau)d\tau \leq C_4. \qquad (4.4.81)$$

Proof. Multiplying (4.4.72), (4.4.73) and (4.4.74) by ε, $\varepsilon^{3/2}$ and ε, respectively; then adding the resultant to (4.4.78), and picking ε small enough, we can obtain the desired estimate (4.4.79).

Multiplying (4.4.51) by $e^{\gamma t}$, using (4.4.49), (4.4.79) and Theorem 4.1.1, and picking $\gamma > 0$ so small that $0 < \gamma \leq \gamma_4 \equiv \min\big[\frac{1}{2C_1}, \gamma_4^{(2)}\big]$, we conclude that for any $t > 0$,

$$e^{\gamma t}\left\|\frac{u_{xxx}}{u}(t)\right\|^2 + (2C_1)^{-1}\int_0^t e^{\gamma \tau}\left\|\frac{u_{xxx}}{u}(\tau)\right\|^2 d\tau$$

$$\leq C_3 + C_1 \int_0^t e^{\gamma \tau}\|E_1(\tau)\|^2 d\tau \leq C_4$$

whence

$$e^{\gamma t}\|u_{xxx}(t)\|^2 + \int_0^t e^{\gamma \tau}\|u_{xxx}(\tau)\|^2 d\tau \leq C_4, \qquad \forall t > 0. \qquad (4.4.82)$$

Similarly to (4.4.53), (4.4.57)–(4.4.58), (4.4.61)–(4.4.62), (4.4.66)–(4.4.67), using (4.4.79) and (4.4.82) and Theorem 4.1.1, we deduce that for any fixed $\gamma \in (0, \gamma_4]$,

$$e^{\gamma t}\Big(\|v_{xxx}(t)\|^2_{H^1} + \|\theta_{xxx}(t)\|^2_{H^1} + \|z_{xxx}(t)\|^2_{H^1}$$

$$+ \|v_{txx}(t)\|^2 + \|\theta_{txx}(t)\|^2 + \|z_{txx}(t)\|^2\Big)$$

$$+ \int_0^t e^{\gamma \tau}\Big(\|v_{xxx}\|^2_{H^1} + \|\theta_{xxx}\|^2_{H^1} + \|z_{xxx}\|^2_{H^1}$$

$$+ \|v_{txx}\|^2 + \|\theta_{txx}\|^2 + \|z_{txx}\|^2\Big)(\tau)d\tau \leq C_4 \qquad (4.4.83)$$

and

$$\int_0^t e^{\gamma\tau}\left(\|v_{tt}\|^2 + \|\theta_{tt}\|^2 + \|z_{tt}\|^2 + \|v_{txxx}\|^2 + \|\theta_{txxx}\|^2 + \|z_{txxx}\|^2\right)(\tau)\,d\tau \le C_4.$$

(4.4.84)

Similarly to (4.4.82), multiplying (4.4.64) by $e^{\gamma t}$, using (4.4.60), (4.4.79), (4.4.82)–(4.4.84) and Theorem 4.1.1, we conclude that for any fixed $\gamma \in (0, \gamma_4]$,

$$e^{\gamma t}\left\|\frac{u_{xxxx}}{u}(t)\right\|^2 + (2C_1)^{-1}\int_0^t e^{\gamma\tau}\left\|\frac{u_{xxxx}}{u}(\tau)\right\|^2 d\tau$$

$$\le C_3 + C_1\int_0^t e^{\gamma\tau}\|E_2(\tau)\|^2\,d\tau \le C_4$$

whence

$$e^{\gamma t}\|u_{xxxx}(t)\|^2 + \int_0^t e^{\gamma\tau}\|u_{xxxx}(\tau)\|^2\,d\tau \le C_4, \quad \forall\, t > 0.$$

(4.4.85)

Similarly to (4.4.68)–(4.4.70), we deduce that for any fixed $\gamma \in (0, \gamma_4]$,

$$\int_0^t e^{\gamma\tau}\left(\|v_{xxxxx}\|^2 + \|\theta_{xxxxx}\|^2 + \|z_{xxxxx}\|^2\right)(\tau)\,d\tau \le C_4, \quad \forall\, t > 0. \qquad (4.4.86)$$

Finally, we derive the desired estimates (4.4.80)–(4.4.81) by (4.4.79) and (4.4.82)–(4.4.86). The proof is complete. □

Proof of Theorem 4.1.2. By Lemmas 4.4.1–4.4.6, Theorem 4.1.1 and Sobolev's embedding theorem, we complete the proof of Theorem 4.1.2. □

Remark 4.4.1. It is easy to find that some proofs of some lemmas in this chapter are applicable to the corresponding lemmas in Chapter 3.

4.5 Bibliographic Comments

In this section, we shall recall some related results. When the material is a polytropic ideal linear viscous gas with constant viscosity and heat conductivity, e.g.,

$$e = C_v\theta, \quad \mathcal{P} = R\frac{\theta}{u} \qquad (4.5.1)$$

with suitable positive constants C_v and R, the global existence and asymptotic behavior of smooth (generalized) solutions to the system (4.1.2)–(4.1.4) with $\delta = 0$ have been investigated by many authors (see, e.g., Kazhikhov [33], Kazhikhov and Shelukhin [36], Nagasawa [45, 46], Okada and Kawashima [47], Qin [59, 52, 55], see also Chapter 3) on the initial boundary value problems and the Cauchy problem. Zheng and Qin [83] established the existence of maximal attractors for the problem (4.1.2)–(4.1.4), (4.1.7) and (4.1.2)–(4.1.8) with $\delta = 0$.

The global existence and asymptotic behavior for the system (4.1.2)–(4.1.4) with $\delta = 0$ (without chemical reactions) have been established only for some constitutive equations and special forms of functions μ, κ and $\mathcal{P}$ before the results of Chapter 3 were published (see [62]). Kazhikhov and Shelukhin [36] proved the global existence and uniqueness of a solution to the initial boundary value problem of a system consisting of a viscous heat-conductive ideal gas bounded by two infinite parallel plates with thermally insulated boundaries, but does not allow heat to be generated by chemical reactions. However, Bebernes and Bressan [2] investigated the case that permits generation by a chemical reaction, so the system contains the chemical species equation (4.1.5). The system (4.1.2)–(4.1.6) was essentially proposed by Kassoy and Poland in [30] for a polytropic ideal gas ($p = 1$), and was studied in Guo and Zhu [21] for global existence, regularity and asymptotic behavior of solutions, in particular, the well-posedness and large time behavior for discontinuous initial data was investigated in Chen, Hoff and Trivisa [5]. In this direction, we would like to mention the works of Qin [57], Qin, Ma, Cavalcanti and Andrade [65], Qin and Muñoz Rivera [66], Qin, Wu and Liu [73], Qin [57, 65, 66, 73] who produced some works on global well-posedness in H^i ($i = 1, 2, 4$) and existence of global attractors for heat-conducting real gas of the compressible Navier-Stokes equations with a constant viscosity coefficient.

As mentioned above, the global existence and the asymptotic behavior in $(H^2[0,1])^4$ of the generalized (global) solutions have never been investigated for equations (4.1.2)–(4.1.4) of the pth power viscous reactive gas with boundary conditions (3.1.7), nor have the existence of classical solutions. Therefore, we continue the work by Lewicka and Mucha [38] ($p \geq 1$) and study the global existence and exponential stability of the solutions in H^i ($i = 2, 4$) in this chapter and have improved the works in [38].

Chapter 5

On a 1D Viscous Reactive and Radiative Gas with First-order Arrhenius Kinetics

5.1 Introduction

In this chapter, we establish the global existence and exponential stability of solutions in H^i $(i = 1, 2, 4)$ for a Stefan-Boltzmann model of a viscous, reactive and radiative gas with first-order Arrhenius kinetics in a bounded interval. In so doing we describe the classical stellar evolution [11] of a finite mass of a heat-conducting viscous reactive fluid in local equilibrium with thermal radiation: pressure, internal energy and thermal conductivity have Stefan-Boltzmann radiative contributions. In order to mimic chemical exchanges inside the fluid, we may consider a simple reacting process with a first-order kinetics, commonly used in combustion theory [12]. The results of this chapter are chosen from [63].

The system under consideration in Lagrangian coordinates reads as follows:

$$u_t = v_x, \tag{5.1.1}$$

$$v_t = \sigma_x, \tag{5.1.2}$$

$$e_t - \sigma v_x + Q_x - \lambda \phi(\theta, Z) = 0, \tag{5.1.3}$$

$$Z_t - \left(\frac{d}{u^2} Z_x \right)_x + \phi(\theta, Z) = 0 \tag{5.1.4}$$

where μ is a constant viscous coefficient, $\lambda \geq 0$ and $d \geq 0$ are two "chemical" constants, and $x \in [0, 1]$ denotes the mass variable, $u(x, t)$ the specific volume, $v(x, t)$ the velocity, $\theta(x, t)$ and $Z(x, t)$ denote the temperature and the fraction of reactant, respectively.

We consider system (5.1.1)–(5.1.4) subject to the boundary conditions

$$v(0, t) = v(1, t) = 0, \ Q(0, t) = Q(1, t) = 0, \ Z_x(0, t) = Z_x(1, t) = 0 \tag{5.1.5}$$

and the initial condition

$$(u, v, \theta, Z)(x, 0) = (u_0, v_0, \theta_0, Z_0)(x). \tag{5.1.6}$$

For the constitutive relations, we consider (see, e.g., [12]) the Stefan-Boltzmann model, i.e., the pressure $p(u, \theta)$, internal energy $e(u, \theta)$, the stress $\sigma(u, v_x, \theta)$ and the thermo-radiative flux $Q(u, \theta)$ take the following forms respectively,

$$p(u, \theta) = R\frac{\theta}{u} + \frac{a}{3}\theta^4, \quad e(u, \theta) = C_v\theta + au\theta^4, \quad \sigma(u, v_x, \theta) = -p(u, \theta) + \mu\frac{v_x}{u}, \tag{5.1.7}$$

$$Q(u, \theta) = -\frac{k(u, \theta)\theta_x}{u}, \qquad k(u, \theta) = \kappa_1 + \kappa_2 u\theta^q \tag{5.1.8}$$

where R, $a, C_v, \mu, \kappa_1, \kappa_2$ and q are positive constants (the values $k_2 = 0$, $a = 0$ correspond to the ideal gas). Finally, the function ϕ in the equations (5.1.3)–(5.1.4) mimics the simplest first-order Arrhenius kinetics (see, e.g., [12]):

$$\phi(\theta, Z) = AZ\theta^\beta e^{-E/(B\theta)} \tag{5.1.9}$$

where A, β, B and E are positive constants.

For a one-dimensional homogeneous real gas, e, σ, η and Q are given by the constitutive relations (see, e.g., [7, 8, 27, 32]):

$$e = e(u, \theta), \quad \sigma = \sigma(u, \theta, v_x), \quad \eta = \eta(u, \theta), \quad Q = Q(u, \theta, \theta_x) \tag{5.1.10}$$

which, in order to be consistent with the Clausius-Duhem inequality (see below (5.1.17)), must satisfy

$$\sigma(u, \theta, 0) = \Psi_u(u, \theta), \quad \eta(u, \theta) = -\Psi_\theta(u, \theta), \tag{5.1.11}$$

$$(\sigma(u, \theta, w) - \sigma(u, \theta, 0))w \geq 0, \quad Q(u, \theta, g)g \leq 0 \tag{5.1.12}$$

where $\Psi = e - \theta\eta$ is the Helmholtz free energy function, and e, σ, η are internal energy, stress and specific entropy, respectively.

For the case of an ideal gas without radiative effect (i.e., $\phi(\theta, Z) = 0$, $Z = 0$),

$$e = C_v\theta, \quad \sigma = -R\frac{\theta}{u} + \mu\frac{v_x}{u}, \quad Q = -\lambda\frac{\theta}{u} \tag{5.1.13}$$

with suitable positive constants C_v, R, μ and λ, the global existence and asymptotic behavior of smooth (generalized) solutions to the system (5.1.1)–(5.1.3) have been investigated by many authors (see, e.g., [33, 36, 45, 46, 47, 49, 50, 59, 73]) on the initial boundary value problems and the Cauchy problem. However, under very high temperature and densities, equation (5.1.13) becomes inadequate. Thus, a more realistic model would be a linearly viscous gas (or Newtonian fluid),

$$\sigma(u, \theta, v_x) = -p(u, \theta) + \frac{\mu(u, \theta)}{u}v_x \tag{5.1.14}$$

satisfying Fourier's law of heat flux,

$$Q(u, \theta, \theta_x) = -\frac{k(u, \theta)}{u}\theta_x \tag{5.1.15}$$

where the internal energy e and the pressure p are coupled by the standard thermodynamical relations

$$e_u(u, \theta) = -p(u, \theta) + \theta p_\theta(u, \theta) \tag{5.1.16}$$

to comply with the Clausius-Duhem inequality,

$$\eta_t + \left(\frac{Q}{\theta}\right)_x \geq 0. \tag{5.1.17}$$

We define three spaces as follows:

$$H_+^1 = \Big\{(u, v, \theta, Z) \in \big(H^1[0, 1]\big)^4 : u(x) > 0, \ \theta(x) > 0, \ 0 \leq Z(x) \leq 1, x \in [0, 1],$$

$$v(0) = v(1) = 0\ \Big\},$$

$$H_+^i = \Big\{(u, v, \theta, Z) \in \big(H^i[0, 1]\big)^4 : u(x) > 0, \ \theta(x) > 0, \ 0 \leq Z(x) \leq 1, x \in [0, 1],$$

$$v(0) = v(1) = 0, \ \ \theta'(0) = \theta'(1) = 0, \ \ Z'(0) = Z'(1) = 0\Big\}, i = 2, 4.$$

Let $E = E_1 \cup E_2$ be a set in the (β, q)-plane in $\mathbb{R}^2$, where

$$E_1 = \Big\{(\beta, q) \in \mathbb{R}^2 : \frac{12}{5} < q, \ 0 < \beta \leq 8\Big\},$$

$$E_2 = \Big\{(\beta, q) \in \mathbb{R}^2 : \frac{5}{2} < q \leq 3, \ \beta < 2q + 6, \ \beta > 8\Big\}$$

$$\bigcup\Big\{(\beta, q) \in \mathbb{R}^2 : 3 \leq q, \ \beta < q + 9, \beta > 8\Big\}.$$

The notation in this chapter is standard. We put $\|\cdot\| = \|\cdot\|_{L^2[0,1]}$ and denote by $C^k(J, B), k \in \mathbb{N}_0$, the space of k-times continuously differentiable functions from $J \subseteq \mathbb{R}$ into a Banach space B, and likewise by $L^p(J, B), (1 \leq p \leq +\infty)$ the corresponding Lebesgue spaces. Subscripts t and x denote the (partial) derivatives with respect to t and x, respectively. We use C_i $(i = 1, 2, 4)$ to denote the generic positive constant depending on the $H^i[0, 1]$ norm of initial datum $(u_0, v_0, \theta_0, Z_0)$, $\min\limits_{x \in [0,1]} u_0(x)$, $\min\limits_{x \in [0,1]} \theta_0(x)$ and $\min\limits_{x \in [0,1]} Z_0(x)$, but independent of variable t. Constant C stands for the absolute positive constant independent of initial data. Without danger of confusion, we shall use the same symbol to denote the state functions as well as their values along a dynamic process, e.g., $p(u, \theta), p(u(x, t), \theta(x, t))$ and $p(x, t)$.

Our main results read as follows.

Theorem 5.1.1. *Let $(\beta, q) \in E$. Assume that the initial data $(u_0, v_0, \theta_0, Z_0) \in H^1_+$ and the compatibility conditions are satisfied, and there exists a constant $\varepsilon_0 > 0$ such that $\bar{u}_0 = \int_0^1 u_0 \, dx \leq \varepsilon_0$, then problem (5.1.1)–(5.1.6) admits a unique global solution $(u(t), v(t), \theta(t), Z(t)) \in H^1_+$ verifying*

$$0 < C_1^{-1} \leq u(x, t) \leq C_1, \quad 0 < C_1^{-1} \leq \theta(x, t) \leq C_1 \text{ on } [0, 1] \times [0, +\infty), \quad (5.1.18)$$

and

$$\|u(t) - \bar{u}\|_{H^1}^2 + \|v(t)\|_{H^1}^2 + \|\theta(t) - \bar{\theta}\|_{H^1}^2 + \|Z(t)\|_{H^1}^2 + \int_0^t \left(\|u - \bar{u}\|_{H^1}^2 + \|v\|_{H^2}^2 \right.$$

$$\left. + \|\theta - \bar{\theta}\|_{H^2}^2 + \|Z\|_{H^2}^2 + \|v_t\|^2 + \|\theta_t\|^2 + \|Z_t\|^2 \right)(\tau) \, d\tau \leq C_1, \quad \forall t > 0$$

$$(5.1.19)$$

where $\bar{u} = \int_0^1 u(x) \, dx = \int_0^1 u_0(x) \, dx$, constant $\bar{\theta} > 0$ is determined by $e(\bar{u}, \bar{\theta}) = \int_0^1 \left(\frac{1}{2} v_0^2 + e(u_0, \theta_0) + \lambda Z_0 \right) dx$.

Moreover, there are constants $C_1 > 0$ and $\gamma_1 = \gamma_1(C_1) > 0$ such that for any fixed $\gamma \in (0, \gamma_1]$, the following estimate holds for any $t > 0$,

$$e^{\gamma t} \left(\|u(t) - \bar{u}\|_{H^1}^2 + \|v(t)\|_{H^1}^2 + \|\theta(t) - \bar{\theta}\|_{H^1}^2 + \|Z(t)\|_{H^1}^2 \right)$$

$$+ \int_0^t e^{\gamma \tau} \left(\|u - \bar{u}\|_{H^1}^2 + \|v\|_{H^2}^2 + \|\theta - \bar{\theta}\|_{H^2}^2 + \|Z\|_{H^2}^2 \right.$$

$$\left. + \|v_t\|^2 + \|\theta_t\|^2 + \|Z_t\|^2 \right)(\tau) \, d\tau \leq C_1. \quad (5.1.20)$$

Theorem 5.1.2. *Let $(\beta, q) \in E$. Assume that the initial data $(u_0, v_0, \theta_0, Z_0) \in H^2_+$ and the compatibility conditions are satisfied, and there exists a constant $\varepsilon_0 > 0$ such that as $\bar{u}_0 = \int_0^1 u_0 \, dx \leq \varepsilon_0$, then there exists a unique global solution $(u(t), v(t), \theta(t), Z(t)) \in H^2_+$ to problem (5.1.1)–(5.1.6) verifying that for any $t > 0$,*

$$\|u(t) - \bar{u}\|_{H^2}^2 + \|v(t)\|_{H^2}^2 + \|\theta(t) - \bar{\theta}\|_{H^2}^2 + \|Z(t)\|_{H^2}^2 + \|v_t(t)\|^2 + \|\theta_t(t)\|^2$$

$$+ \|Z_t(t)\|^2 + \int_0^t \left(\|u - \bar{u}\|_{H^2}^2 + \|v\|_{H^3}^2 + \|\theta - \bar{\theta}\|_{H^3}^2 + \|Z\|_{H^3}^2 \right.$$

$$\left. + \|v_{tx}\|^2 + \|\theta_{tx}\|^2 + \|Z_{tx}\|^2 \right)(\tau) \, d\tau \leq C_2. \quad (5.1.21)$$

Moreover, there are constants $C_2 > 0$ and $\gamma_2 = \gamma_2(C_2) > 0$ such that for any fixed $\gamma \in (0, \gamma_2]$, the following estimate holds for any $t > 0$,

$$e^{\gamma t} \left(\|u(t) - \bar{u}\|_{H^2}^2 + \|v(t)\|_{H^2}^2 + \|\theta(t) - \bar{\theta}\|_{H^2}^2 + \|Z(t)\|_{H^2}^2 + \|v_t(t)\|^2 + \|\theta_t(t)\|^2 \right.$$

$$\left. + \|Z_t(t)\|^2 \right) + \int_0^t e^{\gamma \tau} \left(\|u - \bar{u}\|_{H^2}^2 + \|v\|_{H^3}^2 + \|\theta - \bar{\theta}\|_{H^3}^2 + \|Z\|_{H^3}^2 \right.$$

$$\left. + \|v_{tx}\|^2 + \|\theta_{tx}\|^2 + \|Z_{tx}\|^2 \right)(\tau) \, d\tau \leq C_2. \quad (5.1.22)$$

Theorem 5.1.3. *Let $(\beta, q) \in E$. Assume that the initial data $(u_0, v_0, \theta_0, Z_0) \in H_+^4$ and the compatibility conditions are satisfied, and there exists a constant $\varepsilon_0 > 0$ such that as $\bar{u}_0 = \int_0^1 u_0 \, dx \leq \varepsilon_0$, then there exists a unique global solution $(u(t), v(t), \theta(t), Z(t)) \in H_+^4$ to problem (5.1.1)–(5.1.6) verifying that for any $t > 0$,*

$$\|u(t) - \bar{u}\|_{H^4}^2 + \|v(t)\|_{H^4}^2 + \|\theta(t) - \bar{\theta}\|_{H^4}^2 + \|Z(t)\|_{H^4}^2 + \|v_t(t)\|_{H^2}^2 + \|v_{tt}(t)\|^2$$

$$+ \|\theta_t(t)\|_{H^2}^2 + \|\theta_{tt}(t)\|^2 + \|Z_t(t)\|_{H^2}^2 + \|Z_{tt}(t)\|^2 + \int_0^t \Big(\|u - \bar{u}\|_{H^4}^2 + \|v\|_{H^5}^2$$

$$+ \|\theta - \bar{\theta}\|_{H^5}^2 + \|Z\|_{H^5}^2 + \|v_t\|_{H^3}^2 + \|\theta_t\|_{H^3}^2 + \|Z_t\|_{H^3}^2$$

$$+ \|v_{tt}\|_{H^1}^2 + \|\theta_{tt}\|_{H^1}^2 + \|Z_{tt}\|_{H^1}^2 \Big)(\tau) \, d\tau \leq C_4. \tag{5.1.23}$$

Moreover, there are constants $C_4 > 0$ and $\gamma_4 = \gamma_4(C_4) > 0$ such that for any fixed $\gamma \in (0, \gamma_4]$, the following estimate holds for any $t > 0$,

$$e^{\gamma t} \Big(\|u(t) - \bar{u}\|_{H^4}^2 + \|v(t)\|_{H^4}^2 + \|\theta(t) - \bar{\theta}\|_{H^4}^2 + \|Z(t)\|_{H^4}^2 + \|v_t(t)\|_{H^2}^2 + \|v_{tt}(t)\|^2$$

$$+ \|\theta_t(t)\|_{H^2}^2 + \|\theta_{tt}(t)\|^2 + \|Z_t(t)\|_{H^2}^2 + \|Z_{tt}(t)\|^2 \Big) + \int_0^t e^{\gamma \tau} \Big(\|u - \bar{u}\|_{H^4}^2$$

$$+ \|v\|_{H^5}^2 + \|\theta - \bar{\theta}\|_{H^5}^2 + \|Z\|_{H^5}^2 + \|v_t\|_{H^3}^2 + \|\theta_t\|_{H^3}^2 + \|Z_t\|_{H^3}^2$$

$$+ \|v_{tt}\|_{H^1}^2 + \|\theta_{tt}\|_{H^1}^2 + \|Z_{tt}\|_{H^1}^2 \Big)(\tau) \, d\tau \leq C_4. \tag{5.1.24}$$

Corollary 5.1.1. *Under assumptions of Theorem 5.1.3, the global solution*

$$(u(t), v(t), \theta(t), Z(t)) \in H_+^4$$

obtained in Theorem 5.1.3 is in fact a classical solution verifying for any $\gamma \in (0, \gamma_4]$,

$$\|(u(t) - \bar{u}, v(t), \theta(t) - \bar{\theta}, Z(t))\|_{(C^{3+1/2})^4} \leq C_4 e^{-\gamma t}.$$

Remark 5.1.1. Theorems 5.1.1–5.1.3 still hold for the boundary conditions

$$v(0, t) = v(1, t) = 0, \ \theta(0, t) = \theta(1, t) = T_0 = \text{const.} > 0, \ Z_x(0, t) = Z_x(1, t) = 0$$

where $\bar{\theta}$ should be replaced by T_0.

5.2   Global Existence in H^1

In this section we study the global existence of problem (5.1.1)–(5.1.6) in $H^1[0,1]$ by establishing a series of a priori estimates.

Lemma 5.2.1. *Under the assumptions in Theorem 5.1.1, the following estimates hold:*

$$\int_0^1 Z(x,t)\,dx + \int_0^t \int_0^1 \phi(\theta,Z)(x,s)\,dx ds = \int_0^1 Z_0(x)\,dx, \tag{5.2.1}$$

$$\int_0^1 \left[\frac{1}{2}v^2 + e + \lambda Z\right](x,t)\,dx = \int_0^1 \left[\frac{1}{2}v_0^2 + e_0 + \lambda Z_0\right](x)\,dx = E_0, \tag{5.2.2}$$

$$\theta(x,t) > 0, \quad 0 \le Z(x,t) \le 1, \quad \forall\,(x,t) \in [0,1] \times [0,+\infty), \tag{5.2.3}$$

$$\frac{1}{2}\int_0^1 Z^2(x,t)\,dx + \int_0^t \int_0^1 \frac{d}{u^2}Z_x^2\,dx ds$$

$$+ \int_0^t \int_0^1 Z\phi(\theta,Z)\,dx ds = \frac{1}{2}\int_0^1 Z_0^2(x)\,dx, \tag{5.2.4}$$

$$\Phi(t) + \int_0^t \Psi(s)\,ds \le C_1 \tag{5.2.5}$$

where

$$\Phi(t) = \int_0^1 \left[R(u - \log u - 1) + C_v(\theta - \log\theta - 1)\right](x,t)\,dx,$$

$$\Psi(t) = \int_0^1 \left(\frac{v_x^2}{u\theta} + \frac{k\theta_x^2}{u\theta^2} + \lambda\frac{\phi(\theta,Z)}{\theta}\right)(x,t)\,dx.$$

Proof. Relation (5.2.1) is obtained by integrating (5.1.4) on $[0,1] \times (0,t)$, by using boundary condition (5.1.5). Multiplying by v in (5.1.2), we find the conservation law

$$\left(\frac{1}{2}v^2 + e + \lambda Z\right)_t = \left(\sigma v - Q + \frac{\lambda d}{u^2}Z_x\right)_x. \tag{5.2.6}$$

Now integrating (5.2.6) on $[0,1]$, and using (5.1.1)–(5.1.6), we obtain (5.2.2).

To get the positivity of θ in (5.2.3), we apply the maximum principle to (5.1.3), rewritten as

$$e_\theta \theta_t + \theta p_\theta v_x - \frac{\mu}{u}v_x^2 = \left(\frac{k(u,\theta)}{u}\theta_x\right)_x + \lambda\phi(\theta,Z), \tag{5.2.7}$$

together with (5.1.5), and we use the positivity of θ_0.

To get the positivity of Z, we apply the same principle to the equation (5.1.4), together with (5.1.5), and we use the positivity of Z_0.

To get (5.2.5), we multiply (5.2.7) by θ^{-1},

$$e_\theta \frac{\theta_t}{\theta} + p_\theta u_t = \frac{\mu}{u\theta} v_x^2 + \frac{k(u,\theta)}{u\theta^2}\theta_x^2 + \left(\frac{k(u,\theta)\theta_x}{u\theta}\right)_x + \lambda\frac{\phi(\theta, Z)}{\theta}. \tag{5.2.8}$$

A standard thermodynamical computation of the entropy S takes the form

$$S(u,\theta) = R\log u + C_v \log\theta + \frac{4}{3}au\theta^3 + S_0. \tag{5.2.9}$$

By using the thermodynamical formulation $S_\theta = e_\theta/\theta$, and $S_u = p_\theta$, and by integrating (5.2.8) on $[0,1] \times [0,t]$, we get

$$\int_0^1 \int_0^t \left(\frac{\mu}{u\theta}v_x^2 + \frac{k(u,\theta)}{u\theta^2}\theta_x^2 + \lambda\frac{\phi(\theta, Z)}{\theta}\right) ds dx = \int_0^1 S(x,t)dx - \int_0^1 S_0(x)dx$$

which, with (5.2.9), yields the identity

$$\int_0^t \int_0^1 \left(\frac{\mu}{u\theta}v_x^2 + \frac{k(u,\theta)}{u\theta^2}\theta_x^2 + \lambda\frac{\phi(\theta, Z)}{\theta}\right) dx ds$$

$$+ \int_0^1 (R(u - \log u - 1) + C_v(\theta - \log\theta - 1))\, dx$$

$$= \int_0^1 \left(R(u - 1) + C_v(\theta - 1) + \frac{4}{3}au\theta^3\right) dx.$$

By using estimate (5.1.1) and (5.2.2), we bound the first two terms of the right-hand side. For the last one, we use the Cauchy-Schwartz inequality

$$\int_0^1 u\theta^3 dx \leq \left(\int_0^1 u dx\right)^{1/4} \left(\int_0^1 u\theta^4 dx\right)^{3/4}$$

and we obtain (5.2.5), by using once more (5.2.2).

Relation (5.2.4) is obtained by multiplying the equation (5.1.4) by Z, integrating the result on $[0,1] \times [0,t]$, and using (5.1.5) and (5.2.1). □

The next lemma is a generalization of the Bellman-Gronwall inequality which was established in Qin [52].

Lemma 5.2.2. *Assume that $f(t), g(t)$ and $y(t)$ are nonnegative integrable functions in $[\tau, T]$ $(\tau < T)$ verifying the integral inequality*

$$y(t) \leq g(t) + \int_\tau^t f(s)y(s)ds, \ \ t \in [\tau, T]. \tag{5.2.10}$$

Then we have

$$y(t) \leq g(t) + \int_\tau^t \exp\left(\int_s^t f(\theta)d\theta\right) f(s)g(s)ds, \ \ t \in [\tau, T]. \tag{5.2.11}$$

In addition, if $g(t)$ is a nondecreasing function in $[\tau, T]$, then we conclude

$$y(t) \le g(t) \left[1 + \int_\tau^t \exp\left(\int_s^t f(\theta)d\theta \right) f(s)ds \right], \quad t \in [\tau, T],$$

$$\le g(t) \left[1 + \int_\tau^t f(s)ds \exp\left(\int_\tau^t f(\theta)d\theta \right) \right], \quad t \in [\tau, T]. \tag{5.2.12}$$

If further $T = +\infty$ and $\int_\tau^{+\infty} f(s)ds < +\infty$, then we conclude

$$y(t) \le Cg(t) \tag{5.2.13}$$

where $C = 1 + \int_\tau^{+\infty} f(s)ds \exp\left\{ \int_\tau^{+\infty} f(\theta)d\theta \right\}$ is a positive constant.

Proof. Let

$$G(t) = \int_0^t f(s)y(s)ds. \tag{5.2.14}$$

Differentiating (5.2.14) with respect to t, using (5.2.10), we obtain

$$\frac{d}{dt}G(t) = f(t)y(t) \le f(t)\big(g(t) + G(t)\big). \tag{5.2.15}$$

Thus,

$$\frac{d}{dt}\left(G(t)e^{-\int_0^t f(s)ds} \right) \le f(t)g(t)e^{-\int_0^t f(s)ds}. \tag{5.2.16}$$

Integrating (5.2.16) over (τ, t) with respect to t, we have

$$G(t) \le \int_\tau^t \exp\left(\int_s^t f(\theta)d\theta \right) f(s)g(s)ds$$

which, along with (5.2.10), gives (5.2.11). (5.2.12) and (5.2.13) are obvious. The proof is complete. □

Lemma 5.2.3. *For any $t \ge 0$, there exists one point $x_1 = x_1(t) \in [0,1]$ such that specific volume $u(x,t)$ in problem (5.1.1)–(5.1.6) possesses the following expression: for any $\delta \ge 0$,*

$$u(x,t) = D(x,t)B(t)\left[1 + \mu^{-1}\int_0^t \frac{u(x,s)[p(x,s) - \delta]}{D(x,s)B(s)} ds \right] \tag{5.2.17}$$

where

$$D(x,t) = u_0(x)\exp\left[\mu^{-1}\left(\int_{x_1(t)}^x v(y,t)dy - \int_0^x v_0(y)\, dy \right. \right.$$

$$\left. \left. + \frac{1}{\bar{u}_0(x)}\int_0^1 u_0(x)\int_0^x v_0(y)\, dydx \right) \right], \tag{5.2.18}$$

$$B(t) = \exp\left[-\frac{1}{\mu\bar{u}_0}\int_0^t\int_0^1 (v^2 + up)(x,s)\, dyds + \delta t/\mu \right] \tag{5.2.19}$$

with $\bar{u}_0 = \int_0^1 u_0(x)dx$.

Proof. For $\delta = 0$, (5.2.17) was proved in [1, 3, 33, 36, 45, 46, 47] for a polytropic ideal gas and in [27, 32, 49, 51, 52, 54, 55, 57, 59] for viscous heat-conductive real gas (see also, Lemma 1.2.4). For $\delta > 0$, we shall borrow some ideas from [20, 27, 32, 49, 51, 52, 54, 55, 57, 59]. In fact, for any $\delta \geq 0$, we can rewrite (5.1.1) as

$$u_t = v_x = \frac{1}{\mu}(\sigma + \delta)u + \frac{1}{\mu}u(p - \delta),$$

i.e.,

$$u_t - \frac{1}{\mu}(\sigma + \delta)u = \frac{1}{\mu}u(p - \delta). \tag{5.2.20}$$

Multiplying (5.2.20) by $\exp\{-\frac{1}{\mu}\int_0^t(\sigma + \delta)ds\}$, we infer

$$u(x,t) = \exp\left(\frac{1}{\mu}\int_0^t(\sigma+\delta)ds\right)\left(u_0 + \frac{1}{\mu}\int_0^t u(p-\delta)\exp\left(-\frac{1}{\mu}\int_0^s(\sigma+\delta)d\tau\right)ds\right). \tag{5.2.21}$$

Let

$$h(x, t) = \int_0^t \sigma(x, s)ds + \int_0^x v_0(y)dy. \tag{5.2.22}$$

Then we have

$$h_x = v, \quad h_t = \sigma = \frac{\mu h_{xx}}{u} - p \tag{5.2.23}$$

which, along with (5.1.1) and (5.1.5), implies

$$(uh)_t = hv_x + \mu h_{xx} - pu, \tag{5.2.24}$$

$$h_x|_{x=0,1} = 0. \tag{5.2.25}$$

Integrating (5.2.24) with respect to t and using (5.2.22)–(5.2.25), we get

$$\int_0^1 uhdx = \int_0^1 u_0h_0dx - \int_0^t\int_0^1 (v^2 + up)dxds$$

which, by using the mean value theorem, implies that there exists a point $x_1(t) \in [0, 1]$ for any $t \geq 0$ such that

$$h(x_1(t), t)\int_0^1 udx = \int_0^1 u_0h_0dx - \int_0^t\int_0^1 (v^2 + up)dxds,$$

i.e.,

$$h(x_1(t, t)) = \frac{1}{\bar{u}_0}\left[\int_0^1 u_0h_0dx - \int_0^t\int_0^1 (v^2 + up)dxds\right]. \tag{5.2.26}$$

Thus by (5.2.23)–(5.2.26), we deduce

$$\int_0^t \sigma(x_1(t), s)ds = h(x_1(t), t) - \int_0^{x_1(t)} v_0(y)dy \tag{5.2.27}$$

$$= \frac{1}{\bar{u}_0}\left[\int_0^1 u_0h_0dx - \int_0^t\int_0^1 (v^2 + up)dxds\right] - \int_0^{x_1(t)} v_0(y)dy.$$

Integrating (5.1.2) over $[x_1(t), x] \times [0, t]$ (for any fixed $t \geq 0$), we infer

$$\int_0^t \sigma(x, s)ds = \int_0^t \sigma(x_1(t), s)ds + \int_{x_1(t)}^x (v - v_0)dy. \qquad (5.2.28)$$

Then we infer from (5.2.23), (5.2.27) and (5.2.28) that

$$\int_0^t \sigma(x, s)ds = \frac{1}{\bar{u}_0}\left\{\int_0^1 u_0 h_0 dx - \int_0^t \int_0^1 (v^2 + up)dxds\right\}$$
$$- \int_0^{x_1(t)} v_0(y)dy + \int_{x_1(t)}^x (v - v_0)dy$$
$$= \frac{1}{\bar{u}_0}\left\{\int_0^1 u_0 h_0 dx - \int_0^t \int_0^1 (v^2 + up)dxds\right\}$$
$$+ \int_{x_1(t)}^x v(y, t)dy - \int_0^x v_0(y)dy$$

which, together with (5.2.21), gives (5.2.17). The proof is complete. □

Lemma 5.2.4. *Under the assumptions in Theorem 5.1.1, the following estimate holds:*

$$0 < C_1^{-1} \leq u(x, t) \leq C_1, \quad \forall (x, t) \in [0, 1] \times [0, +\infty). \qquad (5.2.29)$$

Proof. The main idea of the proof is similar to that of Lemma 1.2.5. But the different key point here is to estimate the higher order term $u\theta^4$ which is a contribution of radiative effects from the pressure p and internal energy e (see also (5.1.7)). Thus we need a detailed and careful analysis.

Noting the convexity of function $-\log y$ and using (5.2.5), we have

$$\int_0^1 \theta\, dx - \log \int_0^1 \theta\, dx - 1 \leq \int_0^1 (\theta - \log \theta - 1)\, dx \leq C_1/C_v$$

which implies that there exists a point $b(t) \in [0, 1]$ and two positive constants r_1, r_2 such that

$$0 < r_1 \leq \int_0^1 \theta(x, t)\, dx = \theta(b(t), t) =: \theta_1(t) \leq r_2 \qquad (5.2.30)$$

with r_1, r_2 being two positive roots of equation $y - \log y - 1 = C_1/C_v$. Thus we infer from (5.2.2) that

$$\int_0^1 up\, dx \leq \max[R, a/3] \int_0^1 (\theta + u\theta^4)dx \leq \frac{\max(R, a/3)}{\min(C_v, a)} \int_0^1 ed x \leq \frac{\max[R, a/3]}{\min(C_v, a)}E_0$$

and

$$\int_0^1 v^2 dx \leq 2E_0$$

which, along with (5.2.30), implies

$$0 < a_1 \leq \frac{1}{\mu\bar{u}_0} \int_0^1 (up + v^2)(x, s)\, dx \leq a_2 \tag{5.2.31}$$

with

$$a_1 = Rr_1/\mu\bar{u}_0, \quad a_2 = \left[Rr_2 + (2 + (\max(R, a/3)E_0)/\min(C_v, a))E_0\right]/\mu\bar{u}_0.$$

Using Lemmas 5.2.1–5.2.2, we derive for any $[0,1] \times [0, +\infty)$,

$$0 < C_1^{-1} \leq D(x, t) \leq C_1. \tag{5.2.32}$$

On the other hand, for $0 < m_1 \leq 2$, we infer from Lemma 5.2.1,

$$|\theta^{m_1}(x,t) - \theta^{m_1}(b(t), t)| \leq C_1 \left| \int_{b(t)}^x \theta^{m_1 - 1}\theta_x\, dy \right|$$

$$\leq C_1 \left(\int_0^1 \frac{k(u, \theta)\theta_x^2}{u\theta^2}\, dx \right)^{1/2} \left(\int_0^1 \frac{u\theta^{2m_1}}{k(u, \theta)}\, dx \right)^{1/2}$$

$$\leq C_1 V^{1/2}(t) \left(\int_0^1 \frac{u(1 + \theta^4)}{\kappa_1}\, dx \right)^{1/2}$$

$$\leq C_1 V^{1/2}(t) \tag{5.2.33}$$

where $V(t) = \int_0^1 \frac{k\theta_x^2}{u\theta^2}\, dx$. Thus, for $0 < m_1 \leq 2$,

$$\frac{1}{2}r_1^{2m_1} - C_1 V(t) \leq \theta^{2m_1}(x, t) \leq 2r_2^{2m_1} + C_1 V(t). \tag{5.2.34}$$

Obviously, we derive from Lemmas 5.2.1, 5.2.3 and (5.2.31) for $\delta \geq 0$ and $0 \leq s \leq t$,

$$e^{-(a_2 - \delta/\mu)(t-s)} \leq B(t)B^{-1}(s) \leq e^{-(a_1 - \delta/\mu)(t-s)}. \tag{5.2.35}$$

Thus for $\delta = 0$ in Lemma 5.2.3, we use (5.2.32)–(5.2.35) to derive that there exists a large time $t_0 > 0$ such that, as $t \geq t_0 > 0$,

$$u(x, t) \geq C_1^{-1} \int_0^t \theta e^{-a_2(t-s)}\, ds$$

$$\geq C_1^{-1} \int_0^t \left(\frac{1}{2}r_1^2 - C_1 V(s)\right) e^{-a_2(t-s)}\, ds$$

$$\geq \frac{r_1^2}{2a_2 C_1}(1 - e^{-a_2 t}) - C_1 \int_0^t V(s)e^{-a_2(t-s)}\, ds \geq \frac{r_1^2}{4a_2 C_1} > 0, \tag{5.2.36}$$

where we have used the estimate as $t \to +\infty$,

$$\int_0^t V(s)e^{-a_2(t-s)}\, ds \leq e^{-a_2 t/2} \int_0^\infty V(s)\, ds + \int_{t/2}^t V(s)\, ds \to 0.$$

On the other hand, we can also derive from Lemma 5.2.3 and (5.2.35) (for $s = 0$) that for $\delta = 0$ and for any $t \in [0, t_0]$,

$$u(x,t) \geq \frac{D(x,t)}{B(t)} \geq C_1^{-1} \exp(-a_2 t) \geq C_1^{-1} \exp(-a_2 t_0) > 0$$

which, together with (5.2.36), gives

$$u(x,t) \geq C_1^{-1} > 0, \quad \forall (x,t) \in [0,1] \times [0, +\infty). \tag{5.2.37}$$

Now we begin to show the positive lower bound of u in (5.2.29). To prove it, we need the smallness of the total initial mass $\bar{u}_0$, which is only used to prove the positive lower bound of u.

Now choosing $0 < \varepsilon_0 \leq \frac{3Rr_1}{ar_2^4}$, then as $\bar{u}_0 \leq \varepsilon_0$, we get

$$\frac{Rr_1}{\bar{u}_0} \geq \frac{Rr_1}{\varepsilon_0} > \frac{ar_2^4}{3} \geq \frac{a\theta_1^4}{3}. \tag{5.2.38}$$

Now from (5.2.38), we can pick $\delta > 0$ such that

$$\frac{Rr_1}{\bar{u}_0} \geq \frac{Rr_1}{\varepsilon_0} > \delta > \frac{ar_2^4}{3} \geq \frac{a\theta_1^4}{3}$$

which gives

$$\lambda \equiv a_1 - \frac{\delta}{\mu} = \frac{1}{\mu}[\mu a_1 - \delta] = \frac{1}{\mu}\left[\frac{Rr_1}{\bar{u}_0} - \delta\right] > 0, \tag{5.2.39}$$

$$\delta_1 \equiv \frac{3}{a}\theta_1^{-4}\delta - 1 \geq \frac{3}{a}r_2^{-4}\delta - 1 \equiv \delta_0 > 0. \tag{5.2.40}$$

Noting that by the Young inequality $(b_1 b_2 \leq \varepsilon b_1^2 + b_2^2/(2\varepsilon), \forall \varepsilon, b_1, b_2 > 0)$ and inequality $b^4 - 1 - \varepsilon \leq (b^2 - 1)^2(1 + \varepsilon^{-1})$ with $b \in \mathbb{R}$ and any $\varepsilon > 0$, we derive from (5.2.39)–(5.2.40) that

$$u(p - \delta) = R\theta + u\left(\frac{a}{3}\theta^4 - \delta\right) \tag{5.2.41}$$

and

$$\begin{aligned}
u\left(\frac{a}{3}\theta^4 - \delta\right) &= \frac{a}{3}\theta_1^4 u\left[\tilde{\theta}^4 - \frac{3}{a}\theta_1^{-4}\delta\right] \\
&\leq \frac{a}{3}\theta_1^4 u \max[\tilde{\theta}^4 - 1 - \delta_1, 0] \\
&\leq \frac{a}{3}\theta_1^4 u \max[\tilde{\theta}^4 - 1 - \delta_0, 0] \\
&\leq \frac{a}{3}\theta_1^4 u(\tilde{\theta}^2 - 1)^2(1 + \delta_0^{-1}) \\
&\leq C_1 u(\tilde{\theta}^2 - 1)^2 \tag{5.2.42}
\end{aligned}$$

where

$$\tilde{\theta}(x,t) = \theta(x,t)/\theta_1, \quad \theta_1 \equiv \theta(b(t),t) = \int_0^1 \theta(x,t)\,dx \in [r_1, r_2].$$

Noting that $\tilde{\theta}(b(t),t) = 1$ and by the Poincaré inequality, we deduce

$$(\tilde{\theta}^2 - 1)^2 \le \left(\int_0^1 2|\tilde{\theta}\tilde{\theta}_x|dx\right)^2 \le C_1 \left(\int_0^1 |\theta\theta_x|dx\right)^2$$

$$\le C_1 \left(\int_0^1 \frac{k(u,\theta)\theta_x^2}{u\theta^2}dx\right)\left(\int_0^1 \frac{u\theta^4}{k(u,\theta)}dx\right)$$

$$\le C_1 V(t) \int_0^1 u\theta^4 dx \le C_1 V(t)$$

which, together with (5.2.41)–(5.2.42), gives

$$u(p - \delta) \le R\theta + C_1 V(t)u. \tag{5.2.43}$$

Thus it follows from Lemma 5.2.3, (5.2.32)–(5.2.34), (5.2.42) and (5.2.43) that

$$u(x,t) \le C_1 + C_1 \int_0^t [\theta(x,s) + u(x,s)V(s)] \exp\left(-\lambda(t-s)\right) ds$$

$$\le C_1 + C_1 \int_0^t [1 + V(s) + u(x,s)V(s)] \exp\left(-\lambda(t-s)\right) ds$$

$$\le C_1 + C_1 \int_0^t u(x,s)V(s) \exp\left(-\lambda(t-s)\right) ds$$

that is,

$$F(t) \le C_1 e^{\lambda t} + C_1 \int_0^t V(s)F(s)\,ds \tag{5.2.44}$$

with $F(t) = e^{\lambda t} \max_{x \in [0,1]} u(x,t) \equiv e^{\lambda t} M_u(t)$. Therefore, by Lemma 5.2.2, we deduce

$$F(t) \le C_1 e^{\lambda t} \exp\left(C_1 \int_0^t V(s)ds\right) \le C_1 e^{\lambda t},$$

i.e.,

$$M_u(t) \le C_1$$

which, along with (5.2.37), completes the proof of (5.2.29). $\qquad\square$

Lemma 5.2.5. *Under the assumptions in Theorem 5.1.1, the following estimate holds:*

$$\|u_x(t)\|^2 + \int_0^t \int_0^1 (u_x^2 + \theta u_x^2)(x,s)\,dxds \le C_1 + C_1 A^{q_0}, \quad \forall t > 0 \tag{5.2.45}$$

with $A = \sup_{0 \le s \le t} \|\theta(s)\|_{L^\infty}$ and $q_0 = \max(4 - q, 0)$.

Proof. Equation (5.1.2) can be rewritten as

$$\left(v - \mu \frac{u_x}{u}\right)_t = -p_x. \tag{5.2.46}$$

Multiplying (5.2.46) by $v - \mu \frac{u_x}{u}$ and then integrating the resulting equation over $Q_t \equiv [0,1] \times [0,t]$, we have

$$\frac{1}{2}\left\|v - \mu \frac{u_x}{u}\right\|^2 + R\mu \int_0^t \int_0^1 \frac{\theta u_x^2}{u^3}\, dxds$$

$$= \frac{1}{2}\left\|v_0 - \mu \frac{u_{0x}}{u_0}\right\|^2 + \int_0^t \int_0^1 \left[R\frac{\theta u_x v}{u^2} - \left(R\frac{1}{u} + \frac{4}{3}a\theta^3\right)\theta_x \left(v - \mu \frac{u_x}{u}\right)\right] dxds$$

implying

$$\left\|v - \mu \frac{u_x}{u}\right\|^2 + \int_0^t \int_0^1 \theta u_x^2\, dxds \tag{5.2.47}$$

$$\le C_1 + C_1 \int_0^t \int_0^1 \left[|\theta u_x v| + |\theta_x v| + |\theta^3 \theta_x v| + |\theta_x u_x| + |\theta^3 \theta_x u_x|\right] dxds.$$

Noting the following facts

$$\int_0^t \|v(s)\|_{L^\infty}^2\, ds \le \int_0^t \left(\int_0^1 \frac{v_x^2}{\theta}\, dx\right)\left(\int_0^1 \theta\, dx\right) ds \le C_1 \int_0^t \int_0^1 \frac{v_x^2}{\theta}\, dxds \le C_1,$$

$$\int_0^1 \theta^4\, dx \le C_1,$$

$$\int_0^1 \frac{u\theta^r}{k(u,\theta)}\, dx \le C_1 \int_0^1 (1+\theta)^{r-q}\, dx \le C_1 + C_1 A^{\max(r-q-4,0)}, \quad r \ge 0,$$

and using Lemmas 5.2.1–5.2.4, we easily derive that for any $\varepsilon > 0$,

$$\int_0^t \int_0^1 |\theta u_x v|\, dxds \le \varepsilon \int_0^t \int_0^1 \theta u_x^2\, dxds + C_1(\varepsilon) \int_0^t \|v(s)\|_{L^\infty}^2 \int_0^1 \theta\, dxds$$

$$\le C_1(\varepsilon) + \varepsilon \int_0^t \int_0^1 \theta u_x^2\, dxds, \tag{5.2.48}$$

$$\int_0^t \int_0^1 |\theta_x v|\, dxds \le \left(\int_0^t V(s)ds\right)^{1/2}\left(\int_0^t \int_0^1 \frac{u\theta^2 v^2}{k(u,\theta)}\, dxds\right)^{1/2}$$

$$\le C_1 \left\{\int_0^t \|v(s)\|_{L^\infty}^2 \left(\int_0^1 \theta^2\, dx\right) ds\right\}^{1/2} \le C_1, \tag{5.2.49}$$

$$\int_0^t \int_0^1 |\theta^3 \theta_x v|\, dxds \le \left(\int_0^t V(s)ds\right)^{1/2}\left(\int_0^t \int_0^1 \frac{u\theta^8 v^2}{k(u,\theta)}\, dxds\right)^{1/2}$$

$$\le C_1 + C_1 A^{q_0/2}, \tag{5.2.50}$$

$$\int_0^t \int_0^1 |\theta_x u_x| \, dx ds \leq \left(\int_0^t V(s) ds \right)^{1/2} \left(\int_0^t \int_0^1 \frac{u\theta^2 u_x^2}{k(u,\theta)} dx ds \right)^{1/2}$$

$$\leq C_1 \sup_{0 \leq s \leq t} \left\| \frac{u\theta}{k(u,\theta)} \right\|_{L^\infty}^{1/2} \left(\int_0^t \int_0^1 \theta u_x^2 dx ds \right)^{1/2}$$

$$\leq \varepsilon \int_0^t \int_0^1 \theta u_x^2 dx ds + C_1 A^{\max(1-q,0)} + C_1, \qquad (5.2.51)$$

$$\int_0^t \int_0^1 \left| (\theta^2 - \theta_1^2)\theta\theta_x u_x \right| dx ds$$

$$\leq C_1 \int_0^t \int_0^1 |\theta\theta_x| \, dx \int_0^1 |\theta\theta_x u_x| \, dx ds$$

$$\leq C_1 \sup_{0 \leq s \leq t} \left\{ \left(\int_0^1 \frac{u\theta^4}{k(u,\theta)} dx \right)^{1/2} \left(\int_0^1 \frac{u\theta^4 u_x^2}{k(u,\theta)} dx \right)^{1/2} \right\} \int_0^t V(s) ds$$

$$\leq C_1 + C_1 \sup_{0 \leq s \leq t} \left(\int_0^1 u\theta^4 dx \right)^{1/2} \left(\left\| \frac{u\theta^4}{k(u,\theta)} \right\|_{L^\infty}^{1/2} \|u_x\| \right)$$

$$\leq \varepsilon \sup_{0 \leq s \leq t} \|u_x(s)\|^2 + C_1(\varepsilon) A^{q_0} + C_1, \qquad (5.2.52)$$

$$\int_0^t \int_0^1 |\theta^3 \theta_x u_x| dx ds$$

$$\leq \int_0^t \int_0^1 |(\theta^2 - \theta_1^2)\theta\theta_x u_x| dx ds + \int_0^t \int_0^1 \theta_1^2 |\theta\theta_x u_x| dx ds$$

$$\leq \varepsilon \sup_{0 \leq s \leq t} \|u_x(s)\|^2 + C_1 + C_1 A^{q_0}$$

$$+ C_1 \left(\int_0^t V(s) ds \right)^{1/2} \left(\int_0^t \int_0^1 \frac{u\theta u_x^2}{k(u,\theta)} dx ds \right)^{1/2}$$

$$\leq \varepsilon \sup_{0 \leq s \leq t} \|u_x(s)\|^2 + C_1 + C_1 A^{q_0}$$

$$+ C_1 \sup_{0 \leq s \leq t} \left\| \frac{u\theta^3}{k(u,\theta)} \right\|_{L^\infty}^{1/2} \left(\int_0^t \int_0^1 \theta u_x^2 dx ds \right)^{1/2}$$

$$\leq \varepsilon \sup_{0 \leq s \leq t} \|u_x(s)\|^2 + \varepsilon \int_0^t \int_0^1 \theta u_x^2 dx ds + C_1 + C_1 A^{q_0}. \qquad (5.2.53)$$

Inserting (5.2.48)–(5.2.52) into (5.2.47), we have

$$\|u_x(t)\|^2 + \int_0^t \int_0^1 \theta u_x^2 dx ds \leq C_1 + C_1 A^{q_0} + 3\varepsilon \int_0^t \int_0^1 \theta u_x^2 dx ds + \varepsilon \sup_{0 \leq s \leq t} \|u_x(s)\|^2$$

which, by taking the supremum on the left-hand side and choosing $\varepsilon > 0$ small enough, gives

$$\|u_x(t)\|^2 + \int_0^t \int_0^1 \theta u_x^2 \, dx ds \leq C_1 + C_1 A^{q_0}. \tag{5.2.54}$$

Moreover, noting that, from (5.2.34) for $m_1 = 1/2$,

$$1 \leq C_1 \theta + C_1 V(t), \tag{5.2.55}$$

we derive from (5.2.54)–(5.2.55) that

$$\int_0^t \int_0^1 u_x^2 dx ds \leq C_1 \int_0^t \int_0^1 \theta u_x^2 dx ds + C_1 \int_0^t V(s)\|u_x(s)\|^2 ds$$
$$\leq C_1 + C_1 A^{q_0}$$

which, together with (5.2.54), gives (5.2.44). $\square$

Now repeating the derivation of (5.2.33)–(5.2.34), we conclude

$$\int_0^1 \frac{u\theta^{2m}}{k(u,\theta)} dx \leq C_1 \int_0^1 \frac{\theta^{2m}}{1+\theta^q} dx \leq C_1 \int_0^1 (1+\theta)^{2m-q} dx$$
$$\leq C_1 \int_0^1 (1+\theta^4) dx \leq C_1, \quad 0 \leq m \leq (q+4)/2.$$

We readily obtain the next corollary.

Corollary 5.2.1. *Under the assumptions in Theorem 5.1.1, the following estimate holds:*

$$C_1^{-1} - C_1 V(t) \leq \theta^{2m}(x,t) \leq C_1 + C_1 V(t), \quad 0 \leq m \leq (q+4)/2. \tag{5.2.56}$$

Lemma 5.2.6. *Under the assumptions in Theorem 5.1.1, the following estimates hold:*

$$\int_0^t \int_0^1 (1+\theta)^{2m} u_x^2 \, dx ds \leq C_1 + C_1 A^{q_0}, \quad \forall t > 0, \tag{5.2.57}$$

$$\int_0^t \int_0^1 (1+\theta)^{2m} v^2 \, dx ds \leq C_1, \quad \forall t > 0 \tag{5.2.58}$$

with $0 \leq m \leq (q+4)/2$.

Proof. It follows from Corollary 5.2.1 and Lemma 5.2.5 that

$$\int_0^t \int_0^1 (1+\theta)^{2m} u_x^2 dx ds \leq C_1 \int_0^t \|u_x(s)\|^2 \, ds + \int_0^t V(s)\|u_x(s)\|^2 \, ds \leq C_1 + C_1 A^{q_0}.$$

The proof of (5.2.58) is similar to that of (5.2.57). $\square$

Lemma 5.2.7. *Under the assumptions in Theorem 5.1.1, the following estimates hold for any $t > 0$:*

$$\int_0^t \|v_x(s)\|^2 \, ds \leq C_1 + C_1 A^{q_0}, \tag{5.2.59}$$

$$\|v_x(t)\|^2 + \int_0^t \|v_{xx}(s)\|^2 \, ds \leq C_1 + C_1 A^{q_2}, \tag{5.2.60}$$

$$\int_0^t \|v_t(s)\|^2 \, ds \leq C_1 + C_1 A^{q_3} \tag{5.2.61}$$

where $q_1 = \max(8 - q, 0)$, $q_2 = \max(3q_0, q_1)$, $q_3 = \max[q_1, (q_2 + 3q_0)/2](\geq q_2)$.

Proof. Multiplying (5.1.2) by v, v_{xx} and v_t, respectively, and then integrating the resulting equations over Q_t, using Lemmas 5.2.1–5.2.6, we get

$$\|v(t)\|^2 + \int_0^t \|v_x(s)\|^2 \, ds$$

$$\leq C_1 + C_1 \left| \int_0^t \int_0^1 \left[(1 + \theta^3)\theta_x v + \theta u_x v \right] dx ds \right|$$

$$\leq C_1 + C_1 \int_0^t V(s) \, ds + C_1 \int_0^t \int_0^1 \left(\theta u_x^2 + \theta v^2 + \frac{(1 + \theta^3)^2 \theta^2}{1 + \theta^q} v^2 \right) dx ds$$

$$\leq C_1 + C_1 A^{q_0} + C_1 \int_0^t \|v(s)\|_{L^\infty}^2 \left[\int_0^1 \theta dx + (1 + \|\theta\|_{L^\infty})^{q_0} \int_0^1 (1 + \theta)^4 dx \right] ds$$

$$\leq C_1 + C_1 A^{q_0}, \tag{5.2.62}$$

$$\|v_x(t)\|^2 + \int_0^t \|v_{xx}(s)\|^2 \, ds$$

$$\leq C_1 + C_1 \left| \int_0^t \int_0^1 \left[(1 + \theta^3)\theta_x v_{xx} + \theta u_x v_{xx} + v_x u_x v_{xx} \right] dx ds \right|$$

$$\leq C_1 + \frac{1}{4} \int_0^t \|v_{xx}(s)\|^2 \, ds + C_1 \int_0^t \int_0^1 \left[(1 + \theta^3)^2 \theta_x^2 + v_x^2 u_x^2 + \theta^2 u_x^2 \right] dx ds$$

$$\leq C_1 + \frac{1}{4} \int_0^t \|v_{xx}(s)\|^2 \, ds + C_1 A^{q_0} + C_1(1 + A^{q_0}) \int_0^t \|v_x(s)\|_{L^\infty}^2 \, ds$$

$$\qquad + C_1(1 + A)^{q_1} \int_0^t V(s) \, ds$$

$$\leq C_1 + \frac{1}{4} \int_0^t \|v_{xx}(s)\|^2 \, ds$$

$$\qquad + C_1(1 + A^{q_0}) \left(\int_0^t \|v_x(s)\|^2 \, ds \right)^{1/2} \left(\int_0^t \|v_{xx}(s)\|^2 \, ds \right)^{1/2} + C_1(1 + A)^{q_1}$$

$$\leq C_1 + C_1 A^{3q_0} + C_1 A^{q_1} + \frac{1}{2} \int_0^t \|v_{xx}(s)\|^2 \, ds,$$

i.e.,

$$\|v_x(t)\|^2 + \int_0^t \|v_{xx}(s)\|^2 \, ds \leq C_1(1+A)^{q_2} \tag{5.2.63}$$

with $q_2 = \max(3q_0, q_1)$, and

$$\|v_x(t)\|^2 + \int_0^t \|v_t(s)\|^2 \, ds$$

$$\leq C_1 + C_1 \left| \int_0^t \int_0^1 \left[(1+\theta)^3 \theta_x v_t + \theta u_x v_t + v_x^3 \right] \, dx ds \right|$$

$$\leq C_1 + \frac{1}{2} \int_0^t \|v_t(s)\|^2 \, ds + C_1 \int_0^t \int_0^1 \left[(1+\theta)^6 \theta_x^2 + \theta^2 u_x^2 + |v_x|^3 \right] \, dx ds$$

$$\leq C_1 + \frac{1}{2} \int_0^t \|v_t(s)\|^2 \, ds + C_1(1+A)^{q_1} \int_0^t V(s) \, ds$$

$$\quad + C_1 A^{q_0} + C_1 \int_0^t \|v_x(s)\|^{5/2} \|v_{xx}(s)\|^{1/2} \, ds$$

$$\leq C_1 + C_1 A^{q_0} + C_1 A^{q_1} + \frac{1}{2} \int_0^t \|v_t(s)\|^2 \, ds$$

$$\quad + C_1 \sup_{0 \leq s \leq t} \|v_x(s)\| \left(\int_0^t \|v_x(s)\|^2 \, ds \right)^{3/4} \left(\int_0^t \|v_{xx}(s)\|^2 \, ds \right)^{1/4}$$

$$\leq C_1 + C_1 A^{q_1} + C_1 A^{(q_2+3q_0)/2} + \varepsilon \sup_{0 \leq s \leq t} \|v_x(s)\|^2 + \frac{1}{2} \int_0^t \|v_t(s)\|^2 ds.$$

Thus for sufficiently small $\varepsilon > 0$,

$$\|v_x(t)\|^2 + \int_0^t \|v_t(s)\|^2 \, ds \leq C_1 + C_1 A^{q_3}$$

which, along with (5.2.62)–(5.2.63), yields (5.2.59)–(5.2.61).   □

Corollary 5.2.2. *Under the assumptions in Theorem 5.1.1, the following estimate holds:*

$$\int_0^t \int_0^1 (1+\theta)^{2m} v_x^2 \, dx ds \leq C_1(1+A)^{q_2}, \quad \forall t > 0, \ 0 \leq m \leq (q+4)/2. \tag{5.2.64}$$

Proof. We easily derive from (5.2.56), (5.2.59) and (5.2.60) that

$$\int_0^t \int_0^1 (1+\theta)^{2m} v_x^2 \, dx ds \leq C_1 \int_0^t \int_0^1 v_x^2 dx ds + C_1 \int_0^t \int_0^1 V(s) v_x^2 dx ds \tag{5.2.65}$$

$$\leq C_1 + C_1 A^{q_0} + C_1 A^{q_2} \leq C_1 + C_1 A^{q_2}. \qquad \square$$

Lemma 5.2.8. *Under the assumptions in Theorem 5.1.1, the following estimate holds:*

$$\|(\theta + \theta^4)(t)\|^2 + \int_0^t \int_0^1 (1+\theta)^{q+3} \theta_x^2(x,s)\, dxds \leq C_1 + C_1 A^{q_4}, \quad \forall t > 0 \quad (5.2.66)$$

where

$$q_4 = \begin{cases} \max[q_5, 2q_0, q_0 + 1, (q_0 + q_2)/2] & \text{if } 0 < \beta \leq 8, \\ \max[q_5, 2q_0, q_0 + 1, (q_0 + q_2)/2, 4] & \text{if } \beta > 8 \end{cases}$$

and $q_5 = \max(7 - 2q, 0)$.

Proof. Multiplying (5.1.3) by e and integrating the resulting equation over Q_t, we have

$$\|\theta + \theta^4\|^2 + \int_0^t \int_0^1 (1+\theta)^{q+3} \theta_x^2\, dxds$$

$$\leq C_1 + C_1 \int_0^t \int_0^1 \left|(pe)_x v + ev_x^2 + k(u,\theta)u^{-1}\theta_x\theta^4 u_x + \lambda\phi e\right| dxds$$

$$\leq C_1 + C_1 \int_0^t \int_0^1 \left[(1+\theta)^7|\theta_x v| + (1+\theta)^5|u_x v| + (1+\theta)^{q+4}|\theta_x u_x|\right.$$

$$\left. + (1+\theta)^8|u_x v| + ev_x^2 + |\phi e|\right] dxds. \quad (5.2.67)$$

By Lemmas 5.2.1–5.2.7, we infer

$$\int_0^t \int_0^1 (1+\theta)^7|\theta_x v| dxds$$

$$\leq \varepsilon \int_0^t \int_0^1 (1+\theta)^{q+3}\theta_x^2\, dxds + C_1 \int_0^t \int_0^1 (1+\theta)^{11-q} v^2 dxds$$

$$\leq \varepsilon \int_0^t \int_0^1 (1+\theta)^{q+3}\theta_x^2\, dxds + C_1 + C_1 A^{q_5}, \quad (5.2.68)$$

$$\int_0^t \int_0^1 (1+\theta)^8|u_x v| dxds$$

$$\leq C_1 \int_0^t \int_0^1 (1+\theta)^8 u_x^2\, dxds + C_1 \int_0^t \int_0^1 (1+\theta)^8 v^2\, dxds$$

$$\leq C_1 A^{q_0} \int_0^t \int_0^1 (1+\theta)^{q+4} u_x^2\, dxds + C_1 A^{q_0} \int_0^t \int_0^1 (1+\theta)^{q+4} v^2\, dxds$$

$$\leq C_1 + C_1 A^{2q_0}, \quad (5.2.69)$$

$$\int_0^t \int_0^1 (1+\theta)^{q+4}|\theta_x u_x|\, dxds$$

$$\leq \varepsilon \int_0^t \int_0^1 (1+\theta)^{q+3}\theta_x^2\, dxds + C_1 \int_0^t \int_0^1 (1+\theta)^{q+5} u_x^2\, dxds$$

$$\leq C_1 + C_1 A^{q_0+1} + \varepsilon \int_0^t \int_0^1 (1+\theta)^{q+3}\theta_x^2 dxds, \quad (5.2.70)$$

$$\int_0^t \int_0^1 e v_x^2 \, dx ds$$

$$\leq \int_0^t \|v_x(s)\|_{L^\infty}^2 \left(\int_0^1 e \, dx \right) ds$$

$$\leq C_1 \left(\int_0^t \|v_x(s)\|^2 ds \right)^{1/2} \left(\int_0^t \|v_{xx}(s)\|^2 ds \right)^{1/2}$$

$$\leq C_1 + C_1 A^{(q_0+q_2)/2}, \tag{5.2.71}$$

$$\int_0^t \int_0^1 |\phi e| \, dx ds$$

$$\leq C_1 \int_0^t \int_0^1 (1+\theta^4) \phi \, dx ds \leq C_1 \int_0^t \int_0^1 \phi \, dx ds + C_1 \int_0^t \int_0^1 \theta^4 \phi \, dx ds$$

$$\leq C_1 + C_1 \int_0^t \int_0^1 (1+V(s)) \phi \, dx ds \leq C_1 + C_1 \int_0^t V(s) \left(\int_0^1 \theta^\beta dx \right) ds$$

$$\leq C_1 + C_1 \int_0^t V(s) \left(\int_0^1 \theta^8 dx \right) ds \quad \text{for } 0 < \beta \leq 8 \tag{5.2.72}$$

or

$$\int_0^t \int_0^1 |\phi e| \, dx ds \leq \sup_{0 \leq s \leq t} \|e(s)\|_{L^\infty} \int_0^t \int_0^1 \phi \, dx ds$$

$$\leq C_1 + C_1 A^4 \quad \text{for } \beta > 8. \tag{5.2.73}$$

Thus for $0 < \beta \leq 8$, we infer from (5.2.67)–(5.2.72) that

$$\int_0^1 \theta^8 dx + \int_0^t \int_0^1 (1+\theta)^{q+3} \theta_x^2 \, dx ds$$

$$\leq 2\varepsilon \int_0^t \int_0^1 (1+\theta)^{q+3} \theta_x^2 \, dx ds + C_1 + C_1 A^{q_5} + C_1 A^{2q_0}$$

$$+ C_1 A^{q_0+1} + C_1 A^{(q_0+q_2)/2} + C_1 \int_0^t V(s) \int_0^1 \theta^8 dx ds$$

which, by taking $\varepsilon > 0$ sufficiently small and using the Gronwall inequality, gives (5.2.66). While for $\beta > 8$, taking $\varepsilon > 0$ small enough, we infer from (5.2.67)–(5.2.71) and (5.2.73) that

$$\int_0^1 \theta^8 dx + \int_0^t \int_0^1 (1+\theta)^{q+3} \theta_x^2 \, dx ds$$

$$\leq C_1 + C_1 A^{q_5} + C_1 A^{2q_0} + C_1 A^{q_0+1} + C_1 A^{(q_0+q_2)/2} + C_1 A^4$$

which also implies the desired estimate (5.2.66).   □

Lemma 5.2.9. *Under the assumptions in Theorem 5.1.1, the following estimate holds:*

$$\int_0^1 (1+\theta)^{2q}\theta_x^2\, dx + \int_0^t \int_0^1 (1+\theta)^{q+3}\theta_t^2\, dxds \le C_1(1+A)^{q_{14}}, \quad \forall t>0 \quad (5.2.74)$$

where

$$q_{14} = \max\Big\{q_0 + q_6, q_8, 1 + q_2, q_9 + q_2, 1 + 3(q_0 + q_2)/4,$$

$$(q_4 + q_{10} + q_2)/2, (q_4 + q_{10})/2 + (3q_0 + q_2)/4, q_{12}, q_{13}\Big\},$$

$$q_6 = \max(1 - 2q, 0), \quad q_7 = \max((q - 10)/2, 0),$$

$$q_8 = q_7 + 2q_2, \quad q_9 = \max(1 - q, 0),$$

$$q_{10} = \max(q - 1, 0), \quad q_{11} = \max(q - 3, 0), \quad q'_{11} = \max(3 - q, 0),$$

$$q_{12} = \max\Big\{2q_0 + q_{11} + q_4 + q'_{11}, q_0 + (q_{11} + q_4)/2 + \beta/2,$$

$$q_0 + (q_{11} + q_4)/2 + (3q_2 + q_0)/4\Big\},$$

$$q_{13} = \max[q_{11} + \beta, (\beta + q_2)/2].$$

Proof. Let

$$H(x,t) = H(u,\theta) = \int_0^\theta \frac{k(u,\xi)}{u}\, d\xi = \frac{\kappa_1 \theta}{u} + \frac{\kappa_2 \theta^{q+1}}{q+1}.$$

Then it is easy to verify that

$$H_t = H_u v_x + \frac{k(u,\theta)\theta_t}{u},$$

$$H_{xt} = \left[\frac{k\theta_x}{u}\right]_t + H_u v_{xx} + H_{uu} v_x u_x + \left(\frac{k}{u}\right)_u u_x \theta_t,$$

$$|H_u| + |H_{uu}| \le C_1(1+\theta). \quad (5.2.75)$$

We rewrite (5.1.3) as

$$e_\theta \theta_t + \theta p_\theta v_x - \mu \frac{v_x^2}{u} = \left(\frac{k(u,\theta)}{u}\theta_x\right)_x + \lambda\phi(\theta, Z). \quad (5.2.76)$$

Multiplying (5.2.76) by H_t and integrating the resulting equation over Q_t, we obtain

$$\int_0^t \int_0^1 \left(e_\theta \theta_t + \theta p_\theta v_x - \mu \frac{v_x^2}{u}\right) H_t\, dxds + \int_0^t \int_0^1 \frac{k(u,\theta)}{u}\theta_x H_{tx}\, dxds$$

$$= \int_0^t \int_0^1 \lambda\phi(\theta, Z) H_t\, dxds. \quad (5.2.77)$$

Now we estimate each term in (5.2.77) by using Lemmas 5.2.1–5.2.8 and Corollaries 5.2.1–5.2.2.

We have first

$$\int_0^t \int_0^1 \frac{e_\theta \theta_t^2 k(u,\theta)}{u} \, dxds \geq C_0 \int_0^t \int_0^1 (1+\theta)^{q+3} \theta_t^2 \, dxds. \tag{5.2.78}$$

Now,

$$\left| \int_0^t \int_0^1 e_\theta \theta_t H_u v_x \, dxds \right| \leq C_0 \int_0^t \int_0^1 (1+\theta)^4 |v_x \theta_t| \, dxds$$

$$\leq \frac{C_0}{8} \int_0^t \int_0^1 (1+\theta)^{q+3} \theta_t^2 \, dxds + C_1 \int_0^t \int_0^1 (1+\theta)^{5-q} v_x^2 \, dxds$$

$$\leq \frac{C_0}{8} \int_0^t \int_0^1 (1+\theta)^{q+3} \theta_t^2 \, dxds + C_1(1+A)^{q_6} \int_0^t \int_0^1 (1+\theta)^{q+4} v_x^2 \, dxds$$

$$\leq \frac{C_0}{8} \int_0^t \int_0^1 (1+\theta)^{q+3} \theta_t^2 \, dxds + C_1(1+A)^{q_6+q_0}. \tag{5.2.79}$$

Next,

$$\left| \int_0^t \int_0^1 \left(\theta p_\theta v_x - \mu \frac{v_x^2}{u} \right) \frac{k\theta_t}{u} \, dxds \right|$$

$$\leq C_1 \int_0^t \int_0^1 \left[(1+\theta)^{q+4} |v_x \theta_t| + (1+\theta)^q |\theta_t| v_x^2 \right] dxds$$

$$\leq \frac{C_0}{8} \int_0^t \int_0^1 (1+\theta)^{q+3} \theta_t^2 \, dxds + C_1 \int_0^t \int_0^1 \left[(1+\theta)^{q+5} v_x^2 + (1+\theta)^{q-3} v_x^4 \right] dxds$$

$$\leq \frac{C_0}{8} \int_0^t \int_0^1 (1+\theta)^{q+3} \theta_t^2 \, dxds + C_1(1+A)^{(1+q_2)}$$

$$+ C_1 \int_0^t \int_0^1 (1+\theta)^{q-3} v_x^4 \, dxds. \tag{5.2.80}$$

But

$$\int_0^t \int_0^1 (1+\theta)^{q-3} v_x^4 \, dxds$$

$$\leq C_1(1+A)^{q_7} \int_0^t \int_0^1 (1+\theta)^{(q+4)/2} v_x^4 \, dxds$$

$$\leq C_1(1+A)^{q_7} \left(\int_0^t \int_0^1 v_x^4 \, dxds + \int_0^t V^{\frac{1}{2}}(s) \int_0^1 v_x^4 \, dxds \right)$$

$$\leq C_1(1+A)^{q_7} \left(\int_0^t \|v_x(s)\|^3 \|v_{xx}(s)\| \, ds + \int_0^t V^{\frac{1}{2}}(s) \|v_x(s)\|^3 \|v_{xx}(s)\| \, ds \right)$$

$$\leq C_1(1+A)^{q_7}\left\{\sup_{0\leq s\leq t}\|v_x(s)\|^2\left(\int_0^t\|v_x(s)\|^2\,ds\right)^{\frac{1}{2}}\left(\int_0^t\|v_{xx}(s)\|^2\,ds\right)^{\frac{1}{2}}\right.$$

$$\left.+\sup_{0\leq s\leq t}\|v_x(s)\|^3\left(\int_0^t V(s)\,ds\right)^{\frac{1}{2}}\left(\int_0^t\|v_{xx}(s)\|^2\,ds\right)^{\frac{1}{2}}\right\}$$

$$\leq C_1+C_1A^{q_7+(3q_2+q_0)/2}+C_1A^{2q_2+q_7}$$

$$\leq C_1+C_1A^{2q_2+q_7}\leq C_1+C_1A^{q_8}$$

which, along with (5.2.80), gives

$$\left|\int_0^t\int_0^1\left(\theta p_\theta v_x-\mu\frac{v_x^2}{u}\right)\frac{k\theta_t}{u}\,dxds\right|$$

$$\leq\frac{C_0}{8}\int_0^t\int_0^1(1+\theta)^{q+3}\theta_t^2\,dxds+C_1+C_1A^{q_8}+C_1A^{1+q_2}. \qquad (5.2.81)$$

It follows from Lemma 5.2.7 and Corollary 5.2.2 that

$$\left|\int_0^t\int_0^1(\theta p_\theta v_x-\mu\frac{v_x^2}{u})H_u v_x\,dxds\right|$$

$$\leq C_1\int_0^t\int_0^1\left[(1+\theta)^5 v_x^2+(1+\theta)v_x^3\right]dxds$$

$$\leq C_1(1+A)^{q_9}\int_0^t\int_0^1(1+\theta)^{q+4}v_x^2\,dxds+C_1\int_0^t\int_0^1(1+\theta)|v_x^3|\,dxds$$

$$\leq C_1(1+A)^{q_9+q_2}+C_1(1+A)\int_0^t\int_0^1|v_x|^3\,dxds$$

$$\leq C_1(1+A)^{q_9+q_2}+C_1(1+A)\int_0^t\|v_x(s)\|^{\frac{5}{2}}\|v_{xx}(s)\|^{\frac{1}{2}}\,ds$$

$$\leq C_1(1+A)^{q_9+q_2}+C_1(1+A)\sup_{0\leq s\leq t}\|v_x(s)\|\left(\int_0^t\|v_x(s)\|^2\,ds\right)^{\frac{3}{4}}\left(\int_0^t\|v_{xx}(s)\|^2\,ds\right)^{\frac{1}{4}}$$

$$\leq C_1+C_1A^{q_9+q_2}+C_1A^{1+3(q_0+q_2)/4}. \qquad (5.2.82)$$

Let us consider now the various contributions in the second integral of (5.2.77). By (5.2.64) and Lemmas 5.2.1–5.2.8, we have

$$\int_0^t\int_0^1\frac{k\theta_x}{u}\left(\frac{k\theta_x}{u}\right)_t dxds=\frac{1}{2}\int_0^1\left(\frac{k\theta_x}{u}\right)^2 dx\Big|_0^t$$

$$\geq C_1^{-1}\int_0^1(1+\theta)^{2q}\theta_x^2\,dx-C_1, \qquad (5.2.83)$$

$$\left| \int_0^t \int_0^1 \frac{k\theta_x}{u} H_u v_{xx} \, dxds \right| \leq C_1 \int_0^t \int_0^1 (1+\theta)^{q+1} |\theta_x v_{xx}| \, dxds$$

$$\leq C_1 \left(\int_0^t \int_0^1 (1+\theta)^{q+3} \theta_x^2 \, dxds \right)^{\frac{1}{2}} \left(\int_0^t \int_0^1 (1+\theta)^{q-1} v_{xx}^2 \, dxds \right)^{\frac{1}{2}}$$

$$\leq C_1 (1+A)^{(q_4+q_{10}+q_2)/2} \tag{5.2.84}$$

and

$$\left| \int_0^t \int_0^1 \frac{k\theta_x}{u} H_{uu} v_x u_x \, dxds \right| \leq C_1 \int_0^t \int_0^1 (1+\theta)^{q+1} |\theta_x u_x v_x| \, dxds$$

$$\leq C_1 \left(\int_0^t \int_0^1 (1+\theta)^{q+3} \theta_x^2 \, dxds \right)^{\frac{1}{2}} \left(\int_0^t \int_0^1 (1+\theta)^{q-1} v_x^2 u_x^2 \, dxds \right)^{\frac{1}{2}}$$

$$\leq C_1 (1+A)^{(q_4+q_{10})/2} \left(\int_0^t \|v_x(s)\|_{L^\infty}^2 \|u_x(s)\|^2 \, ds \right)^{\frac{1}{2}}$$

$$\leq C_1 (1+A)^{(q_4+q_{10})/2+(3q_0+q_2)/4}. \tag{5.2.85}$$

Noting the following facts

$$\int_0^t \int_0^1 v_x^4 \, dxds \leq C_1 \int_0^t \|v_x(s)\|^3 \|v_{xx}(s)\| \, ds$$

$$\leq C_1 \sup_{0\leq s\leq t} \|v_x(s)\|^2 \left(\int_0^t \|v_x(s)\|^2 \, ds \right)^{\frac{1}{2}} \left(\int_0^t \|v_{xx}(s)\|^2 \, ds \right)^{\frac{1}{2}}$$

$$\leq C_1 (1+A)^{(3q_2+q_0)/2},$$

$$\int_0^t \int_0^1 \phi^2 \, dxds \leq C_1 A^\beta \int_0^t \int_0^1 Z\phi \, dxds \leq C_1 + C_1 A^\beta$$

and using equation (5.1.3), we obtain

$$\int_0^t \left\| \left(\frac{k(u,\theta)\theta_x}{u} \right)_x \right\|^2 ds \leq C_1 \int_0^t \left(\|e_\theta \theta_t\|^2 + \|\theta p_\theta v_x\|^2 + \|v_x^2\|^2 + \|\phi\|^2 \right)(s) ds$$

$$\leq C_1 \int_0^t \int_0^1 \left[(1+\theta)^6 \theta_t^2 + (1+\theta)^8 v_x^2 + v_x^4 + Z\theta^\beta \phi \right] dxds$$

$$\leq C_1 (1+A)^{q'_{11}} \int_0^t \int_0^1 (1+\theta)^{q+3} \theta_t^2 \, dxds + C_1 (1+A)^{q_0} \int_0^t \int_0^1 (1+\theta)^{q+4} v_x^2 \, dxds$$

$$+ C_1 \sup_{0\leq s\leq t} \|v_x(s)\|^2 \left(\int_0^t \|v_x(s)\|^2 ds \right)^{1/2} \left(\int_0^t \|v_{xx}(s)\|^2 ds \right)^{1/2} + C_1 (1+A)^\beta$$

$$\leq C_1 (1+A)^{q'_{11}} \int_0^t \int_0^1 (1+\theta)^{q+3} \theta_t^2 \, dxds + C_1 + C_1 A^{(q_0+3q_2)/2} + C_1 A^\beta. \tag{5.2.86}$$

Thus, by the Sobolev inequality, Lemmas 5.2.5 and 5.2.8, we conclude

$$\left| \int_0^t \int_0^1 \frac{k(u,\theta)\theta_x}{u} \cdot \left(\frac{k(u,\theta)}{u}\right)_u u_x \theta_t \, dx ds \right| \le C_1 \int_0^t \int_0^1 (1+\theta)^q |\theta_x u_x \theta_t| \, dx ds$$

$$\le \frac{C_0}{8} \int_0^t \int_0^1 (1+\theta)^{q+3} \theta_t^2 \, dx ds + C_1 \int_0^t \int_0^1 \frac{(1+\theta)^{q-3}}{k^2(u,\theta)} \left(\frac{k\theta_x}{u}\right)^2 u_x^2 \, dx ds$$

$$\le \frac{C_0}{8} \int_0^t \int_0^1 (1+\theta)^{q+3} \theta_t^2 \, dx ds + C_1 \int_0^t \left\| \frac{k\theta_x}{u} \right\|_{L^\infty}^2 \|u_x\|^2 \, ds$$

$$\le \frac{C_0}{8} \int_0^t \int_0^1 (1+\theta)^{q+3} \theta_t^2 \, dx ds + C_1(1+A)^{q_0} \int_0^t \left\| \frac{k\theta_x}{u} \right\| \left\| \left(\frac{k\theta_x}{u}\right)_x \right\| \, ds$$

$$\le \frac{C_0}{8} \int_0^t \int_0^1 (1+\theta)^{q+3} \theta_t^2 \, dx ds$$
$$+ C_1(1+A)^{q_0} \left(\int_0^t \left\| \frac{k\theta_x}{u} \right\|^2 ds\right)^{\frac{1}{2}} \left(\int_0^t \left\| \left(\frac{k\theta_x}{u}\right)_x \right\|^2 ds\right)^{\frac{1}{2}}$$

$$\le \frac{C_0}{8} \int_0^t \int_0^1 (1+\theta)^{q+3} \theta_t^2 \, dx ds$$
$$+ C_1(1+A)^{q_{11}/2+q_0} \left(\int_0^t \int_0^1 (1+\theta)^{q+3} \theta_x^2 \, dx ds\right)^{\frac{1}{2}} \left(\int_0^t \left\| \left(\frac{k\theta_x}{u}\right)_x \right\|^2 ds\right)^{\frac{1}{2}}$$

$$\le \frac{C_0}{4} \int_0^t \int_0^1 (1+\theta)^{q+3} \theta_t^2 \, dx ds + C_1 + C_1 A^{2q_0+q_{11}+q_4+q'_{11}}$$
$$+ C_1 A^{q_0+q_{11}/2+q_4/2+\beta/2} + C_1 A^{(3q_2+q_0)/4+q_0+q_{11}/2+q_4/2}$$

$$\le \frac{C_0}{4} \int_0^t \int_0^1 (1+\theta)^{q+3} \theta_t^2 \, dx ds + C_1 + C_1 A^{q_{12}}. \tag{5.2.87}$$

The last contribution is

$$\left| \int_0^t \int_0^1 \lambda\phi\left(H_u v_x + \frac{k}{u}\theta_t\right) dx ds \right| \le C_1 \int_0^t \int_0^1 \left[(1+\theta)|v_x\phi| + (1+\theta)^q|\theta_t\phi|\right] dx ds$$

$$\le \frac{C_0}{8} \int_0^t \int_0^1 (1+\theta)^{q+3} \theta_t^2 \, dx ds + C_1 \int_0^t \int_0^1 (1+\theta)^{q-3} \phi^2 \, dx ds$$
$$+ C_1 \left(\int_0^t \int_0^1 \phi^2 \, dx ds\right)^{1/2} \left(\int_0^t \int_0^1 (1+\theta)^2 v_x^2 \, dx ds\right)^{1/2}$$

$$\le \frac{C_0}{8} \int_0^t \int_0^1 (1+\theta)^{q+3} \theta_t^2 \, dx ds + C_1 + C_1 A^{q_{11}+\beta} + C_1 A^{(\beta+q_2)/2}$$

$$\le \frac{C_0}{8} \int_0^t \int_0^1 (1+\theta)^{q+3} \theta_t^2 \, dx ds + C_1(1+A)^{q_{13}}. \tag{5.2.88}$$

Therefore estimate (5.2.74) follows from (5.2.77)–(5.2.88). □

Lemma 5.2.10. *Under the assumptions in Theorem 5.1.1, the following estimates hold:*

$$\|\theta(t)\|_{L^\infty} \le C_1, \quad \forall t > 0, \tag{5.2.89}$$

$$\|u_x(t)\|^2 + \|v_x(t)\|^2 + \|\theta_x(t)\|^2 + \|Z_x(t)\|^2 \tag{5.2.90}$$

$$+ \int_0^t \Big(\|u_x\|^2 + \|v_x\|^2 + \|\theta_x\|^2 + \|Z_x\|^2 + \|v_t\|^2$$

$$+ \|v_{xx}\|^2 + \|\theta_t\|^2 + \|\theta_{xx}\|^2 + \|Z_t\|^2 + \|Z_{xx}\|^2 \Big)(s)ds \le C_1, \quad \forall t > 0.$$

Proof. By Lemmas 5.2.1–5.2.9 and the Young inequality, we infer

$$\left| \theta^{q+5} - \theta_1^{q+5} \right| \le C_1 \int_0^1 |\theta^{q+4}\theta_x| \, dx$$

$$\le C_1 \left(\int_0^1 \theta^{2q}\theta_x^2 \, dx \right)^{1/2} \left(\int_0^1 \theta^8 \, dx \right)^{1/2}$$

$$\le C_1(1 + A)^{(q_{14}+q_4)/2}$$

which gives

$$A^{q+5} \le C_1 + C_1 A^{(q_{14}+q_4)/2} \le \frac{1}{2}A^{q+5} + C_1 \tag{5.2.91}$$

where we have used that $(\beta, q) \in E$ implies $q_{14} + q_4 < 2q + 10$. Therefore, we derive from (5.2.91) and the Young inequality that

$$A \le C_1, \quad \|\theta(t)\|_{L^\infty} \le C_1$$

which, together with Lemmas 5.2.5–5.2.9, gives

$$\|u_x(t)\|^2 + \|v_x(t)\|^2 + \|\theta_x(t)\|^2 + \int_0^t \left(\|u_x\|^2 + \|v_x\|^2 \right.$$

$$+ \|\theta_x\|^2 + +\|v_{xx}\|^2 + \|v_t\|^2 + \|\theta_t\|^2 \big)(s)ds \le C_1, \quad \forall t > 0. \tag{5.2.92}$$

Multiplying (5.1.4) by Z_{xx}, Z_t, respectively, and then integrating the resulting equations over $[0, 1] \times [0, t]$, using (5.2.92), Lemmas 5.2.1 and 5.2.4, we obtain

$$\|Z_x(t)\|^2 + \int_0^t \|Z_{xx}(s)\|^2 \, ds$$

$$\le C_1 + \frac{1}{4} \int_0^t \|Z_{xx}(s)\|^2 \, ds + C_1 \int_0^t \int_0^1 (u_x^2 Z_x^2 + \phi^2)(x, s) \, dx ds$$

$$\le C_1 + \frac{1}{4} \int_0^t \|Z_{xx}(s)\|^2 \, ds + C_1 \int_0^t \|Z_x(s)\|_{L^\infty}^2 \, dx ds$$

$$\le C_1 + \frac{1}{2} \int_0^t \|Z_{xx}(s)\|^2 \, ds,$$

i.e.,

$$\|Z_x(t)\|^2 + \int_0^t \|Z_{xx}(s)\|^2\, ds \le C_1 \qquad (5.2.93)$$

and

$$\|Z_x(t)\|^2 + \int_0^t \|Z_t(s)\|^2\, ds \le C_1 + C_1 \int_0^t \int_0^1 \left[v_x^2 + Z_x^4 + \phi^2 \right] dx ds$$

$$\le C_1 + C_1 \int_0^t \|Z_x(s)\|_{L^\infty}^2 \|Z_x(s)\|^2\, ds$$

$$\le C_1 + C_1 \int_0^t \|Z_x(s)\|^2\, ds + C_1 \int_0^t \|Z_{xx}(s)\|^2\, ds \le C_1$$

which, along with (5.2.92)–(5.2.93), implies (5.2.90). $\qquad\square$

Remark 5.2.1. We also use the following estimate to derive $A \le C_1$.

$$\left| \theta^{q+3} - \theta_1^{q+3} \right| \le C_1 \int_0^1 |\theta^{q+2}\theta_x|\, dx$$

$$\le C_1 \left(\int_0^1 \theta^{2q}\theta_x^2\, dx \right)^{1/2} \left(\int_0^1 \theta^4\, dx \right)^{1/2}$$

$$\le C_1(1+A)^{q_{14}/2}$$

which implies

$$A^{q+3} \le C_1 + C_1 A^{q_{14}/2} \le \frac{1}{2}A^{q+3} + C_1,$$

i.e.,

$$A \le C_1,$$

provided that $q_{14} < 2q + 6$. However, after a lengthy calculation we easily know that the range of (β, q) derived by $q_{14} < 2q + 6$ is smaller than that derived by $q_{14} + q_4 < 2q + 10$. More precisely,

$$(\beta, q) \in D \equiv \left\{ (\beta, q) \in \mathbb{R}^2 : \beta > 0, q > 0, q_{14} < 2q + 6 \right\}$$

$$= D_1 \cup (D_2 \cap D_3 \cap D_4)$$

$$\subset E \equiv (E_1 \cup E_2) = \left\{ (\beta, q) \in \mathbb{R}^2 : \beta > 0, q > 0, q_{14} + q_4 < 2q + 10 \right\}$$

where

$$D_1 = \left\{ (\beta, q) \in \mathbb{R}^2 : \frac{5}{2} < q, 0 < \beta \le 8 \right\},$$

$$D_2 = \left\{ (\beta, q) \in \mathbb{R}^2 : \frac{5}{2} < q \le 3, 8 < \beta < 6q \right\} \bigcup \left\{ (\beta, q) \in \mathbb{R}^2 : 3 < q \le 4, 8 < \beta < 5q+3 \right\}$$

$$\bigcup \left\{ (\beta, q) \in \mathbb{R}^2 : 4 < q, 8 < \beta < 3q + 11 \right\},$$

$$D_3 = \left\{ (\beta,q) \in \mathbb{R}^2 : \frac{5}{2} < q \leq 3, 8 < \beta < 2q+6 \right\} \bigcup \left\{ (\beta,q) \in \mathbb{R}^2 : 3 \leq q, 8 < \beta < q+9 \right\},$$

$$D_4 = \left\{ (\beta,q) \in \mathbb{R}^2 : \frac{5}{2} < q \leq 8, 8 < \beta < 5q+4 \right\} \bigcup \left\{ (\beta,q) \in \mathbb{R}^2 : 8 \leq q, 8 < \beta < 4q+2 \right\}.$$

Lemma 5.2.11. *Under the assumptions in Theorem 5.1.1, the following estimates hold for any $t > 0$:*

$$\frac{d}{dt} \|u_x(t)\|^2 \leq C_1 (\|u_x(t)\|^2 + \|v_{xx}(t)\|^2), \tag{5.2.94}$$

$$\frac{d}{dt} \|v_x(t)\|^2 \leq C_1 (\|\theta_x(t)\|^2 + \|u_x(t)\|^2 + \|v_x(t)\|^2 + \|v_{xx}(t)\|^2), \tag{5.2.95}$$

$$\frac{d}{dt} \|\theta_x(t)\|^2 + \int_0^1 (1+\theta)^{q-3} \theta_{xx}^2 \, dx$$
$$\leq C_1 (\|v_x(t)\|^2 + \|v_{xx}(t)\|^2 + \|\theta_x(t)\|^2 + \|\phi(t)\|^2), \tag{5.2.96}$$

$$\frac{d}{dt} \|Z_x(t)\|^2 \leq C_1 (\|Z_x(t)\|^2 + \|Z_{xx}(t)\|^2 + \|\phi(t)\|^2). \tag{5.2.97}$$

Proof. Differentiating (5.1.1) with respect to x and multiplying the resulting equation by u_x, we get estimate (5.2.94).

Similarly, multiplying (5.1.2) and (5.1.4) by v_{xx}, Z_{xx}, respectively, and then integrating the resulting equation on $[0,1]$, we obtain estimates (5.2.95) and (5.2.97).

Multiplying (5.2.76) by $e_\theta^{-1} \theta_{xx}$ and integrating the resulting equation on $[0,1]$, we deduce

$$\frac{d}{dt} \|\theta_x(t)\|^2 + 2 \int_0^1 \frac{k\theta_{xx}^2}{ue_\theta} \, dx = \int_0^1 \left(\frac{\theta p_\theta v_x}{e_\theta} - \mu \frac{v_x^2}{ue_\theta} - \frac{k_x \theta_x}{ue_\theta} + \frac{k\theta_x u_x \theta_{xx}}{u^2 e_\theta} + \frac{\phi}{e_\theta} \right) \theta_{xx} \, dx$$

$$\leq \varepsilon \|\theta_{xx}\|^2 + C_1 \left(\|v_x\|^2 + \|v_x\|_{L^4}^4 + \|\theta_x\|_{L^4}^4 + \|u_x \theta_x\|^2 + \|\phi\|^2 \right)$$

$$\leq \varepsilon \|\theta_{xx}\|^2 + C_1 \left(\|v_x\|^2 + \|v_x\|^3 \|v_{xx}\| + \|\theta_x\|^3 \|\theta_{xx}\| + \|\theta_x\|^4 + \|\theta_x\|_{L^\infty}^2 + \|\phi\|^2 \right)$$

$$\leq 2\varepsilon \int_0^1 \frac{k\theta_{xx}^2}{ue_\theta} \, dx + C_1 (\|v_x\|^2 + \|v_{xx}\|^2 + \|\theta_x\|^2 + \|\phi\|^2).$$

Taking $\varepsilon > 0$ sufficiently small in the above inequality, we obtain (5.2.96). □

Lemma 5.2.12. *Under the assumptions in Theorem 5.1.1, the following estimates hold as $t \to +\infty$:*

$$\|u(t) - \bar{u}\|_{H^1} \to 0, \quad \|v(t)\| \to 0, \quad \|v(t)\|_{L^\infty} \to 0, \tag{5.2.98}$$

$$\|\theta_x(t)\| \to 0, \quad \|\theta(t) - \bar{\theta}\|_{H^1} \to 0, \quad \|\theta(t) - \bar{\theta}\|_{L^\infty} \to 0, \tag{5.2.99}$$

$$\|Z(t)\|_{H^1} \to 0, \tag{5.2.100}$$

$$0 < C_1^{-1} \leq \theta(x,t) \leq C_1, \quad \forall (x,t) \in [0,1] \times [0,+\infty). \tag{5.2.101}$$

Proof. By using Lemmas 5.2.1–5.2.11, we know

$$\int_0^t (\|u_x\|^2 + \|v_x\|^2 + \|\theta_x\|^2 + \|Z_x\|^2)(s)\, ds \leq C_1 \tag{5.2.102}$$

and

$$\int_0^t \left(\left|\frac{d}{dt}\|u_x\|^2\right| + \left|\frac{d}{dt}\|v_x\|^2\right| + \left|\frac{d}{dt}\|\theta_x\|^2\right| + \left|\frac{d}{dt}\|Z_x\|^2\right| \right)(s)\, ds \leq C_1 \tag{5.2.103}$$

which conclude (5.2.98)–(5.2.100).

We derive from (5.2.99) that there exists a large time $t_0 > 0$ such that for any $t \geq t_0$,

$$\theta(x,t) \geq \frac{1}{2}\bar{\theta} > 0. \tag{5.2.104}$$

On the other hand, if we put $\Theta := \frac{1}{\theta}$, then (5.2.76) becomes

$$e_\theta \Theta_t = \left(\frac{k}{u}\Theta_x\right)_x + \frac{up_\theta^2}{4\mu} - \left[\frac{2k\Theta_x^2}{u\Theta} + \mu\frac{\Theta^2}{u}\left(v_x - \frac{up_\theta}{2\mu\Theta}\right)^2 + \lambda\Theta^2\phi \right]$$

which, together with (5.2.29) and (5.2.89), implies that there exists a positive constant C_1 such that

$$\Theta_t \leq \frac{1}{e_\theta}\left(\frac{k}{u}\Theta_x\right)_x + C_1.$$

Defining $\widetilde{\Theta}(x,t) := C_1 t + \max\limits_{[0,1]} \frac{1}{\theta_0(x)} - \Theta(x,t)$ and a parabolic operator $\mathcal{L} := -\frac{\partial}{\partial t} + \frac{1}{e_\theta}\frac{\partial}{\partial x}\left(\frac{k}{u}\frac{\partial}{\partial x}\right)$, we have a system

$$\begin{cases} \mathcal{L}\widetilde{\Theta} \leq 0, & \text{on } Q_{t_0}, \\ \widetilde{\Theta}|_{t=0} \geq 0, & \text{on } [0,1], \\ \widetilde{\Theta}_x|_{x=0,1} = 0, & \text{on } [0,t_0]. \end{cases}$$

The standard comparison argument implies

$$\min_{(x,t)\in\bar{Q}_{t_0}} \widetilde{\Theta}(x,t) \geq 0$$

which gives for any $(x,t) \in \bar{Q}_{t_0}$,

$$\theta(x,t) \geq \left(C_1 t + \max_{x\in[0,1]} \frac{1}{\theta_0(x)} \right)^{-1}.$$

Thus,

$$\theta(x,t) \geq \left(C_1 t_0 + \max_{x\in[0,1]} \frac{1}{\theta_0(x)} \right)^{-1} \geq C_1^{-1}, \quad 0 \leq t \leq t_0$$

which, together with (5.2.104) and (5.2.89), gives (5.2.101). $\qquad\square$

5.3 Exponential Stability in H^1

In this section, we shall establish the exponential stability of solutions in H^1_+.
Let $\rho = \frac{1}{u}$, and we easily get that the specific entropy

$$\eta = \eta(u, \theta) = \eta(\rho, \theta) = R \log u + \frac{4a}{3} u\theta^3 + C_v \log \theta \qquad (5.3.1)$$

satisfies

$$\frac{\partial \eta}{\partial \rho} = -\frac{p_\theta}{\rho^2}, \quad \frac{\partial \eta}{\partial \theta} = \frac{e_\theta}{\theta} \qquad (5.3.2)$$

with $p = R\rho\theta + \frac{a}{3}\theta^4$.

We consider the transform

$$\mathcal{A} : (\rho, \theta) \in \mathcal{D}_{\rho,\theta} = \{(\rho, \theta) : \rho > 0,\ \theta > 0\} \to (u, \eta) \in \mathcal{AD}_{\rho,\theta},$$

where $u = 1/\rho$ and $\eta = \eta(1/\rho, \theta)$. Owing to the Jacobian

$$\frac{\partial(u, \eta)}{\partial(\rho, \theta)} = -\frac{e_\theta}{\rho^2 \theta} = -\frac{1}{\rho^2} \left(C_v \theta^{-1} + 4a\rho^{-1}\theta^2 \right) < 0, \qquad (5.3.3)$$

there is a unique inverse function $(u, \eta) \in \mathcal{AD}_{\rho,\theta}$. Thus the function e, p can be also regarded as the smooth functions of (u, η). We write

$$e = e(u, \eta) :\equiv e(u, \theta(u, \eta)) = e(1/\rho, \theta),$$
$$p = p(u, \eta) :\equiv p(u, \theta(u, \eta)) = p(1/\rho, \theta).$$

Let

$$\mathcal{E}(u, v, \eta, Z) = \frac{v^2}{2} + \lambda Z + e(u, \eta) - e(\bar{u}, \bar{\eta}) - \frac{\partial e}{\partial u}(\bar{u}, \bar{\eta})(u - \bar{u}) - \frac{\partial e}{\partial \eta}(\bar{u}, \bar{\eta})(\eta - \bar{\eta}), \quad (5.3.4)$$

where $e(u, \eta) = e(u, \theta) = C_v \theta + a\rho^{-1}\theta^4$, $\bar{u} = \int_0^1 u_0\, dx$ and $\bar{\theta} > 0$ is determined by

$$e(\bar{u}, \bar{\theta}) = e(\bar{u}, \bar{\eta}) = \int_0^1 \left(\frac{1}{2}v_0^2 + e(u_0, \theta_0) + \lambda Z_0 \right) dx \qquad (5.3.5)$$

and

$$\bar{\eta} = \eta(\bar{u}, \bar{\theta}). \qquad (5.3.6)$$

Lemma 5.3.1. *Under assumptions of Theorem 5.1.1, there holds that, for any* $(x, t) \in [0, 1] \times [0, +\infty)$,

$$\frac{v^2}{2} + \lambda Z + C_1^{-1} \left(|u - \bar{u}|^2 + |\eta - \bar{\eta}|^2 \right) \leq \mathcal{E}(u, v, \eta, Z)$$

$$\leq \frac{v^2}{2} + \lambda Z + C_1 \left(|u - \bar{u}|^2 + |\eta - \bar{\eta}|^2 \right). \qquad (5.3.7)$$

Proof. The proof is similar to that of Lemmas 3.2.3 and 4.3.1, we omit the detail here. □

Lemma 5.3.2. *Under assumptions of Theorem 5.1.1, for any $(u_0, v_0, \theta_0, Z_0) \in H^1_+$ there are positive constants $C_1 > 0$ and $\gamma'_1 = \gamma'_1(C_1) < \gamma_0/2$ such that for any fixed $\gamma \in (0, \gamma'_1]$, there holds for any $t > 0$,*

$$
e^{\gamma t}\Big(\|v(t)\|^2 + \|Z(t)\|_{L^1(0,1)} + \|Z(t)\|^2 + \|u(t) - \bar{u}\|^2
$$
$$
+ \|\rho(t) - \bar{\rho}\|^2 + \|\eta(t) - \bar{\eta}\|^2 + \|\rho_x(t)\|^2 + \|u_x(t)\|^2 \Big)
$$
$$
+ \int_0^t e^{\gamma s} \Big(\|Z_x\|^2 + \|\theta_x\|^2 + \|v_x\|^2
$$
$$
+ \int_0^1 \phi(\theta, Z)\, dx + \|u_x\|^2 + \|\rho_x\|^2 \Big)(s)\, ds \le C_1. \tag{5.3.8}
$$

Proof. The proof is similar to that of Lemma 4.3.2. □

Lemma 5.3.3. *Under assumptions of Theorem 5.1.1, for any $(u_0, v_0, \theta_0, Z_0) \in H^1_+$, there are positive constants $C_1 > 0$ and $\gamma_1 = \gamma_1(C_1) \le \gamma'_1$ such that for any fixed $\gamma \in (0, \gamma_1]$, the following estimate holds:*

$$
e^{\gamma t} \big(\|v_x(t)\|^2 + \|\theta_x(t)\|^2 + \|Z_x(t)\|^2 \big)
$$
$$
+ \int_0^t e^{\gamma s} \Big(\|v_{xx}\|^2 + \|\theta_{xx}\|^2 + \|Z_{xx}\|^2
$$
$$
+ \|v_t\|^2 + \|\theta_t\|^2 + \|Z_t\|^2 \Big)(s)\, ds \le C_1, \quad \forall t > 0. \tag{5.3.9}
$$

Proof. The proof is similar to that of Lemma 4.3.3. □

5.4 Proof of Theorem 5.1.2

In this section, we shall sketch the proof of Theorem 5.1.2 just by giving a series of lemmas whose proofs will be omitted since similar arguments to those in Chapter 4 and [55] can be used.

Lemma 5.4.1. *Under assumptions of Theorem 5.1.2, for any $(u_0, v_0, \theta_0, Z_0) \in H^2_+$, we have, for any $t > 0$,*

$$
\|v_{xx}(t)\|^2 + \|v_x(t)\|^2_{L^\infty} + \|v_t(t)\|^2 + \int_0^t \|v_{tx}(\tau)\|^2\, d\tau \le C_2, \tag{5.4.1}
$$

$$
\|\theta_{xx}(t)\|^2 + \|\theta_x(t)\|^2_{L^\infty} + \|\theta_t(t)\|^2 + \int_0^t \|\theta_{tx}(\tau)\|^2\, d\tau \le C_2. \tag{5.4.2}
$$

Proof. We refer to Lemma 2.3.1 and Lemma 4.3.4 or see, e.g., Lemma 3.2 in [55]. □

Lemma 5.4.2. *Under assumptions of Theorem 5.1.2, for any $(u_0, v_0, \theta_0, Z_0) \in H_+^2$ we have, for any $t > 0$,*

$$\|u_{xx}(t)\|^2 + \|u_x(t)\|_{L^\infty}^2 + \int_0^t \|u_{xx}(\tau)\|^2 \, d\tau \leq C_2, \tag{5.4.3}$$

$$\|Z_{xx}(t)\|^2 + \|Z_x(t)\|_{L^\infty}^2 + \|Z_t(t)\|^2 + \int_0^t \|Z_{tx}(\tau)\|^2 \, d\tau \leq C_2, \tag{5.4.4}$$

$$\int_0^t (\|v_{xxx}\|^2 + \|\theta_{xxx}\|^2 + \|Z_{xxx}\|^2)(\tau) \, d\tau \leq C_2. \tag{5.4.5}$$

Proof. We refer to the proofs of Lemmas 4.3.4–4.3.5. □

Lemma 5.4.3. *Under assumptions of Theorem 5.1.2, for any $(u_0, v_0, \theta_0, Z_0) \in H_+^2$, there exists a positive constant $\gamma_2' = \gamma_2'(C_2) \leq \gamma_1$ such that for any fixed $\gamma \in (0, \gamma_2']$, the generalized solution $(u(t), v(t), \theta(t), Z(t))$ in H_+^2 to the problem (5.1.1)–(5.1.6) satisfies the following estimate:*

$$e^{\gamma t} \left(\|v_t(t)\|^2 + \|\theta_t(t)\|^2 + \|Z_t(t)\|^2 + \|v_{xx}(t)\|^2 + \|\theta_{xx}(t)\|^2 + \|Z_{xx}(t)\|^2 \right)$$

$$+ \int_0^t (\|v_{tx}\|^2 + \|\theta_{tx}\|^2 + \|Z_{tx}\|^2)(\tau) \, d\tau \leq C_2, \quad \forall t > 0. \tag{5.4.6}$$

Proof. We refer to the proofs of Lemmas 4.3.4–4.3.5. □

Lemma 5.4.4. *Under assumptions of Theorem 5.1.2, for any $(u_0, v_0, \theta_0, Z_0) \in H_+^2$, there exists a positive constant $\gamma_2 = \gamma_2(C_2) \leq \gamma_2'$ such that for any fixed $\gamma \in (0, \gamma_2]$, the generalized solution $(u(t), v(t), \theta(t), Z(t))$ in H_+^2 to the problem (5.1.1)–(5.1.6) satisfies the following estimate for any $t > 0$,*

$$e^{\gamma t} \|u_{xx}(t)\|^2 + \int_0^t e^{\gamma \tau} (\|u_{xx}\|^2 + \|v_{xxx}\|^2 + \|\theta_{xxx}\|^2 + \|Z_{xxx}\|^2)(\tau) \, d\tau \leq C_2. \tag{5.4.7}$$

Proof. We refer to the proof of Lemma 4.3.5. □

Proof of Theorem 5.1.2. By Lemmas 5.4.1–5.4.4 and Theorem 5.1.1, we complete the proof of Theorem 5.1.2. □

5.5   Proof of Theorem 5.1.3

In this section, we shall sketch the proof of Theorem 5.1.3 just by giving a series of lemmas whose proofs will be omitted since similar arguments to those in Chapter 4 and [57] can be used.

Lemma 5.5.1. *Under assumptions of Theorem 5.1.3, for any* $(u_0, v_0, \theta_0, Z_0) \in H_+^4$, *we have for* ε *small enough,*

$$\|v_{tx}(x,0)\| + \|\theta_{tx}(x,0)\| + \|Z_{tx}(x,0)\| \leq C_3, \tag{5.5.1}$$

$$\begin{aligned} \|v_{tt}(x,0)\| + \|\theta_{tt}(x,0)\| + \|Z_{tt}(x,0)\| + \|v_{txx}(x,0)\| \\ + \|\theta_{txx}(x,0)\| + \|Z_{txx}(x,0)\| \leq C_4, \end{aligned} \tag{5.5.2}$$

$$\|v_{tt}(t)\|^2 + \int_0^t \|v_{ttx}(\tau)\|^2 \, d\tau \leq C_4 + C_4 \int_0^t \|\theta_{txx}(\tau)\|^2 \, d\tau, \quad \forall t > 0, \tag{5.5.3}$$

$$\begin{aligned} \|\theta_{tt}(t)\|^2 + \int_0^t \|\theta_{ttx}(\tau)\|^2 \, d\tau \leq C_4 + C_2\varepsilon^{-1} \int_0^t (\|\theta_{txx}\|^2 + \|Z_{txx}\|^2)(\tau) \, d\tau \\ + C_1\varepsilon \int_0^t (\|v_{ttx}\|^2 + \|v_{txx}\|^2)(\tau) \, d\tau, \quad \forall t > 0, \end{aligned} \tag{5.5.4}$$

$$\|Z_{tt}(t)\|^2 + \int_0^t \|Z_{ttx}(\tau)\|^2 \, d\tau \leq C_4 + C_4 \int_0^t (\|\theta_{txx}\|^2 + \|Z_{txx}\|^2)(\tau) \, d\tau, \quad \forall t > 0. \tag{5.5.5}$$

Proof. We refer to the proof of Lemma 4.4.1.   □

Lemma 5.5.2. *Under assumptions of Theorem 5.1.3, for any* $(u_0, v_0, \theta_0, Z_0) \in H_+^4$, *the following estimates hold for any* $t > 0$ *and for* $\varepsilon \in (0,1)$ *small enough,*

$$\begin{aligned} \|v_{tx}(t)\|^2 + \int_0^t \|v_{txx}(\tau)\|^2 \, d\tau \\ \leq C_4 + C_1\varepsilon^2 \int_0^t (\|v_{ttx}\|^2 + \|\theta_{txx}\|^2)(\tau) \, d\tau, \end{aligned} \tag{5.5.6}$$

$$\begin{aligned} \|\theta_{tx}(t)\|^2 + \int_0^t \|\theta_{txx}(\tau)\|^2 \, d\tau \\ \leq C_4 + C_2\varepsilon^2 \int_0^t (\|\theta_{ttx}\|^2 + \|v_{txx}\|^2 + \|\theta_{tx}\|^2 \|\theta_{xxx}\|^2)(\tau) \, d\tau, \end{aligned} \tag{5.5.7}$$

$$\|Z_{tx}(t)\|^2 + \int_0^t \|Z_{txx}(\tau)\|^2 \, d\tau \leq C_4. \tag{5.5.8}$$

Proof. We refer to the proof of Lemma 4.4.2.   □

Lemma 5.5.3. *Under assumptions of Theorem 5.1.3, for any $(u_0, v_0, \theta_0, Z_0) \in H_+^4$, we have for any $t > 0$,*

$$\|v_{tt}(t)\|^2 + \|v_{tx}(t)\|^2 + \|\theta_{tt}(t)\|^2 + \|\theta_{tx}(t)\|^2 + \|Z_{tt}(t)\|^2 + \|Z_{tx}(t)\|^2$$

$$+ \int_0^t \Big(\|v_{ttx}\|^2 + \|v_{txx}\|^2 + \|\theta_{ttx}\|^2$$

$$+ \|\theta_{txx}\|^2 + \|Z_{ttx}\|^2 + \|Z_{txx}\|^2 \Big)(\tau) \, d\tau \leq C_4, \tag{5.5.9}$$

$$\|v_{xxx}(t)\|_{H^1}^2 + \|\theta_{xxx}(t)\|_{H^1}^2 + \|Z_{xxx}(t)\|_{H^1}^2 + \|v_{txx}(t)\|^2 + \|\theta_{txx}(t)\|^2 + \|Z_{txx}(t)\|^2$$

$$+ \int_0^t \Big(\|v_{tt}\|^2 + \|\theta_{tt}\|^2 + \|Z_{tt}\|^2$$

$$+ \|v_{txx}\|_{H^1}^2 + \|\theta_{txx}\|_{H^1}^2 + \|Z_{txx}\|_{H^1}^2 \Big)(\tau) \, d\tau \leq C_4, \tag{5.5.10}$$

$$\int_0^t \Big(\|v_{xxxx}\|_{H^1}^2 + \|\theta_{xxxx}\|_{H^1}^2 + \|Z_{xxxx}\|_{H^1}^2 \Big)(\tau) \, d\tau \leq C_4. \tag{5.5.11}$$

Proof. We refer to the proofs of Lemmas 4.4.2–4.4.3. $\qquad\square$

Lemma 5.5.4. *Under assumptions of Theorem 5.1.3, for any $(u_0, v_0, \theta_0, Z_0) \in H_+^4$, there exists a positive constant $\gamma_4^{(1)} = \gamma_4^{(1)}(C_4) \leq \gamma_2(C_2)$ such that for any fixed $\gamma \in (0, \gamma_4^{(1)}]$, the following estimates hold for any $t > 0$ and $\varepsilon \in (0,1)$ small enough:*

$$e^{\gamma t}\|v_{tt}(t)\|^2 + \int_0^t e^{\gamma\tau}\|v_{ttx}(\tau)\|^2 \, d\tau \leq C_4 + C_2 \int_0^t e^{\gamma\tau}\|\theta_{txx}(\tau)\|^2 \, d\tau, \tag{5.5.12}$$

$$e^{\gamma t}\|\theta_{tt}(t)\|^2 + \int_0^t e^{\gamma\tau}\|\theta_{ttx}(\tau)\|^2 \, d\tau$$

$$\leq C_4\varepsilon^{-3} + C_2\varepsilon^{-1} \int_0^t e^{\gamma\tau}(\|\theta_{txx}\|^2 + \|Z_{txx}\|^2)(\tau) \, d\tau$$

$$+ \varepsilon \int_0^t e^{\gamma\tau}(\|v_{txx}\|^2 + \|v_{ttx}\|^2)(\tau) \, d\tau, \tag{5.5.13}$$

$$e^{\gamma t}\|Z_{tt}(t)\|^2 + \int_0^t e^{\gamma\tau}\|Z_{ttx}(\tau)\|^2 \, d\tau \leq C_4 + C_2 \int_0^t e^{\gamma\tau}(\|\theta_{txx}\|^2 + \|Z_{txx}\|^2)(\tau) \, d\tau. \tag{5.5.14}$$

Proof. We refer to the proof of Lemma 4.4.4. $\qquad\square$

Lemma 5.5.5. *Under assumptions of Theorem 5.1.3, for any $(u_0, v_0, \theta_0, Z_0) \in H_+^4$, there exists a positive constant $\gamma_4^{(2)} \leq \gamma_4^{(1)}$ such that for any fixed $\gamma \in (0, \gamma_4^{(2)}]$, the following estimates hold for any $t > 0$ and $\varepsilon \in (0,1)$ small enough:*

$$e^{\gamma t}\|v_{tx}(t)\|^2 + \int_0^t e^{\gamma\tau}\|v_{txx}(\tau)\|^2 \, d\tau \leq C_4 + C_2\varepsilon^2 \int_0^t e^{\gamma\tau}(\|v_{ttx}\|^2 + \|\theta_{txx}\|^2)(\tau) \, d\tau, \tag{5.5.15}$$

$$e^{\gamma t}\|\theta_{tx}(t)\|^2 + \int_0^t e^{\gamma \tau}\|\theta_{txx}(\tau)\|^2\, d\tau \le C_4 + \varepsilon^2 \int_0^t e^{\gamma \tau}(\|\theta_{ttx}\|^2 + \|v_{txx}\|^2)(\tau)\, d\tau,$$

$$\tag{5.5.16}$$

$$e^{\gamma t}\|Z_{tx}(t)\|^2 + \int_0^t e^{\gamma \tau}\|Z_{txx}(\tau)\|^2\, d\tau \le C_4, \tag{5.5.17}$$

$$e^{\gamma t}(\|v_{tx}(t)\|^2 + \|\theta_{tx}(t)\|^2 + \|Z_{tx}(t)\|^2) + \int_0^t e^{\gamma \tau}(\|v_{txx}\|^2 + \|\theta_{txx}\|^2 + \|Z_{txx}\|^2)(\tau)\, d\tau$$

$$\le C_4 + C_2\varepsilon^2 \int_0^t e^{\gamma \tau}(\|v_{ttx}\|^2 + \|\theta_{ttx}\|^2)(\tau)\, d\tau. \tag{5.5.18}$$

Proof. We refer to the proof of Lemma 4.4.5. □

Lemma 5.5.6. *Under assumptions of Theorem 5.1.3, for any $(u_0, v_0, \theta_0, Z_0) \in H_+^4$, there is a positive constant $\gamma_4 \le \gamma_4^{(2)}$ such that for any fixed $\gamma \in (0, \gamma_4]$, the following estimates hold for any $t > 0$,*

$$e^{\gamma t}\left(\|v_{tt}(t)\|^2 + \|\theta_{tt}(t)\|^2 + \|Z_{tt}(t)\|^2 + \|v_{tx}(t)\|^2 + \|\theta_{tx}(t)\|^2 + \|Z_{tx}(t)\|^2\right)$$

$$+ \int_0^t e^{\gamma \tau}\left(\|v_{ttx}\|^2\|\theta_{ttx}\|^2 + \|Z_{ttx}\|^2 + \|v_{txx}\|^2 + \|\theta_{txx}\|^2 + \|Z_{txx}\|^2\right)(\tau)\, d\tau \le C_4,$$

$$\tag{5.5.19}$$

$$e^{\gamma t}\|u_{xxx}(t)\|_{H^1}^2 + \int_0^t e^{\gamma \tau}\|u_{xxx}(\tau)\|_{H^1}^2\, d\tau \le C_4, \tag{5.5.20}$$

$$e^{\gamma t}\left(\|v_{xxx}(t)\|_{H^1}^2 + \|\theta_{xxx}(t)\|_{H^1}^2 + \|Z_{xxx}(t)\|_{H^1}^2\right.$$

$$\left. + \|v_{txx}(t)\|^2 + \|\theta_{txx}(t)\|^2 + \|Z_{txx}(t)\|^2\right)$$

$$+ \int_0^t e^{\gamma \tau}\left(\|v_{xxxx}\|_{H^1}^2 + \|\theta_{xxxx}\|_{H^1}^2 + \|Z_{xxxx}\|_{H^1}^2 + \|v_{txx}\|_{H^1}^2 + \|\theta_{txx}\|_{H^1}^2\right.$$

$$\left. + \|Z_{txx}\|_{H^1}^2 + \|v_{tt}\|^2 + \|\theta_{tt}\|^2 + \|Z_{tt}\|^2\right)(\tau)\, d\tau \le C_4. \tag{5.5.21}$$

Proof. We refer to the proof of Lemma 4.4.6. □

Proof of Theorem 5.1.3. By Lemmas 5.5.1–5.5.6 and Theorems 5.1.1–5.1.2, we complete the proof of Theorem 5.1.3. □

5.6 Bibliographic Comments

Radiation hydrodynamics (see, e.g., Mihalas and Weibel-Mihalas [44], Pomraning [48], Williams [77]) describes the propagation of thermal radiation through a fluid or gas. Similarly to ordinary fluid mechanics, the equations of motion are derived from conservation laws for macroscopic quantities. However, when radiation is present, the classical "material" flow has to be coupled with the radiation which is an assembly of photons and needs a priori a relativistic treatment (the photons are massless particles travelling at the speed of light). The whole problem under consideration when the matter is in local thermodynamical equilibrium (LTE) is thus a coupling between standard hydrodynamics for the matter and a radiative transfer equation for the photon distribution, through a suitable description, such as in plasma when the radiation is LTE with matter and velocities are not too large, a non-relativistic one temperature description is possible [44, 77]. Moreover, if the matter is extremely radiative opaque, so that the matter free-path of photons is much smaller than the typical length of the flow, we obtain a simplified description (radiation hydrodynamics in the diffusion limit) which amounts to solving a standard hydrodynamical (compressible Navier-Stokes) problem system with additional correction terms in the pressure, the internal energy and the thermal conduction. To describe richer physical processes, for simplicity we may consider the fluid as reactive and couple the dynamics with the first-order chemical kinetics of combustion type, namely the one-order Arrhenius kinetics. Although it is simplified, this model can be proved to model correctly some astrophysical situations of interest such as stellar evolution or interstellar medium dynamics (see, e.g., Chen [3], Ducomet [11]).

For a viscous heat-conducting real gas with (5.1.10) and (5.1.14)–(5.1.16) without radiative effect ($Z \equiv 0, \phi(\theta, Z) \equiv 0$), there are many results on the global existence and large-time behavior of solutions; we refer to Hsiao and Luo [24], Luo [39], Matsumara and Nishida [43], Kawohl [32], Jiang [26, 27] and Qin [49, 50, 51, 52, 54, 55, 57, 59], etc.

In recent years, heat-conducting radiative viscous gas has drawn the attention of a number of mathematicians (see, e.g., Donatelli and Triosia [10], Ducomet [11, 12], Ducomet, Feireisl, Petzeltov and Straškraba [13], Ducomet and Feireisl [14, 15], Ducomet and Zlotnik [16, 17, 18, 19], Secchi [74], Umehara and Tani [75]). Among them, we would like to mention the work by Ducomet [12] with constitutive relations (5.1.14)–(5.1.16) in which the global existence and exponential decay in H^1 of smooth solutions to the 1D viscous reactive and radiative gas for problem (5.1.1)–(5.1.6) were established. However, we should note that there indeed exist some defects and even mistakes in the argument in [12], for example, in Lemmas 2-3, 13, etc. More precisely, first, in Lemma 2 (concerning the representation of the specific volume u), the term $\bar{u}_0 = \int_0^1 u_0(x)dx$ in $D(x,t)$ and $B(t)$ was missed (cf., Lemma 5.2.3 in this chapter); second, in the derivation of (19) in Lemma 3 in [12], we only know the estimate $\int_0^1 u\theta^4 dx \leq C = \text{const.} > 0$ due to Lemma 5.2.1; the author used the wrong estimate $\int_0^1 \theta^4 dx \leq C = \text{const.} > 0$, so the range of q ($2r - q = 4$) in the derivation of (19) in [12] is indeed wrong; third, in the proof of positive lower and upper uniform bounds of the specific volume u on page 1340 in Lemma 13 in [12], the wrong estimate $up \leq C(1+\theta^4)$ was used, while the correct estimate should be $up \leq C(\theta+u\theta^4)$, this estimate is very crucial for the proof of the positive upper bound of u (see, e.g., Lemma 5.2.4); fourth, the proof in [12] of Theorem 3, which describes the exponential decay of global solutions in H^1 for a large time, was not given, in fact its detailed proof does not merely follow as the author indicated in [12]; for a $1D$ viscous heat-conducting real gas, some new techniques are needed. Just for the above reasons, it is necessary for us to rephrase first the detailed proof of the positive lower and upper bounds of the specific volume u in Lemma 5.2.4, and then to give a detailed proof of Theorem 5.1.1. To prove Lemma 5.2.4, we have to modify the representation of the specific volume u in Lemma 5.2.3 (in fact, the proof in [12] has some defects, we have to rephrase Lemma 5.2.3 by introducing a parameter $\delta \geq 0$). On the other hand, Theorem 5.1.1 has improved Theorem 3 in [12] which only stated that the global solutions exponentially decay for a large time, that is, our results in Theorem 5.1.1 indicate the global solutions are exponentially stable for all $t > 0$ (not merely for a large time). Moreover, the range of $(\beta, q) \in \{q \geq 4, \beta > 0\}$ was considered in [12], while the larger range of $(\beta, q) \in E$ (see Theorem 5.1.1) is considered in this chapter. We should also note that the exponential decay of $\|Z\|$ (see (5.3.8)) plays a crucial role in establishing exponential stability of solutions in H^1 (see the proof of Lemma 5.3.2).

In this chapter, we first correct some defects in [12] and establish the existence of global solution in $H^i(i = 1, 2, 4)$ for problem (5.1.1)–(5.1.6) with constitutive relations (5.1.7)–(5.1.8). Note that the ranges of exponents q and r in Qin [51, 52, 54] cover those of q and $r = 3$ in (5.1.7)–(5.1.8); however, $p(u, \theta)$, $e(u, \theta)$ and $k(u, \theta)$ in (5.1.7)–(5.1.8) can not satisfy assumptions in Qin [51, 52, 54] completely. Therefore it is necessary for us to investigate problem (5.1.1)–(5.1.6).

There are two difficulties to overcome in proving our results. The first arises from the higher order nonlinearities of temperature θ in $p(u, \theta)$, $e(u, \theta)$ and $k(u, \theta)$ in (5.1.7)–(5.1.8). To overcome this difficulty, we make use of Corollary 5.2.1 and the interpolation techniques to reduce the higher order of θ. The second is that in order to study the exponential stability, we need to establish uniform estimates depending only on the initial data, but independent of any length of time.

The main novelties of this chapter are as follows:

(1) We have corrected some defects in [12] and improved the results in [12] for the global solutions in H^1.
(2) We bound the norms of (u, v, θ, Z) and their derivatives in terms of an expression of the form $(1 + \sup_{0 \leq s \leq t} \|\theta\|_{L^\infty})^\Lambda$ with Λ being a positive constant only depending on q and β.
(3) We first establish the global existence and exponential stability of global solutions in H^2 and H^4 for boundary condition (5.1.5) for this model.
(4) We first establish the global existence and exponential stability of classical solutions for boundary condition (5.1.5) for this model.

Bibliography

[1] S.N. Antontsev, A.V. Kazhikhov and V.N. Monakhov, Boundary Value Problems in Mechanics of Nonhomogeneous Fluids, Amsterdam, New York, 1990.

[2] J. Bebernes and A. Bressan, Global a priori estimates for a viscous reactive gas, Proc. Roy. Soc. Edinburgh, Sect. A, **101**(1985), 321–333.

[3] G. Chen, Global solutions to the compressible Navier-Stokes equations for a reacting mixture, SIAM J. Math. Anal., **23**(1992), 609–634.

[4] G. Chen, D. Hoff and K. Trivisa, Global solutions to the compressible Navier-Stokes equations with large discontinuous initial data, Comm. Partial Differential Equations, **25**(2000), 2233–2257.

[5] G. Chen, D. Hoff and K. Trivisa, On the Navier-Stokes equations for exothermically reacting compressible fluids, Acta Math. Appl. Sinica, **18**(2002), 1–8.

[6] Y. Cho, H. Choe, and H. Kim, Unique solvability of the initial boundary value problems for compressible viscous fluids, J. Math. Pures Appl., **83**(2004), 243–275.

[7] C.M. Dafermos, Global smooth solutions to the initial boundary value problem for the equations for one-dimensional nonlinear thermoviscoelasticity, SIAM J. Math. Anal., **13**(1982), 397–408.

[8] C.M. Dafermos and L. Hsiao, Global smooth thermomechnical processes in one-dimensional nonlinear thermoviscoelasticity, Nonlinear Anal., TMA, **6**(1982), 435–454.

[9] K. Deckelnick, L^2 Decay for the compressible Navier-Stokes equations in unbounded domains, Comm. Partial Differential Equations, **18**(1993), 1445–1476.

[10] D. Donatelli and K. Triosia, On the motion of a viscous compressible radiative-reacting gas, Comm. Math. Phys., **265**(2006), 463–491.

[11] B. Ducomet, Hydrodynamical models of gaseous stars, Reviews Math. Phys., **8**(1996), 957–1000.

[12] B. Ducomet, A model of thermal dissipaption for a one-dimensional viscous reactive and radiative gas, Math. Meth. Appl. Sci., **22**(1999), 1323–1349.

[13] B. Ducomet, E. Feireisl, H. Petzeltov and I. Straškraba, Existence globale pour un fluide barotrope autogravitant, C. R. Acad. Sci. Paris, Série I, **332**(2001), 627–632.

[14] B. Ducomet and E. Feireisl, On the dynamics of gaseous stars, Arch. Rat. Mech. Anal., **174**(2004), 221–266.

[15] B. Ducomet and E. Feireisl, The equations of magnetohydrodynamics: on the interation between matter and radiation in the evolution of gaseous stars, Comm. Math. Phys., **266**(2006), 595–626.

[16] B. Ducomet and A.A. Zlotnik, Stabilization for viscous compressible heat-conducting media equations with nonmonotone state functions, C. R. Acad. Sci. Paris, Ser. I, **334**(2002),119–124.

[17] B. Ducomet and A.A. Zlotnik, Stabilization for equations of one-dimensional viscous compressible heat-conducting media with nonmonotone equation of state, J. Differential Equations, **194**(2003), 51–81.

[18] B. Ducomet and A.A. Zlotnik, Stabilization for 1D radiative and reactive viscous gas flows, C. R. Acad. Sci. Paris, Ser. I, **338**(2004), 127–132.

[19] B. Ducomet and A. Zlotnik, On the large-time behavior of 1D radiative and reactive viscous flows for higher-order kinetics, Nonlinear Anal., TMA, **63**(2005), 1011–1033.

[20] B. Ducomet and A.A. Zlotnik, Lyapunov functional method for 1D radiative and reactive viscous gas dynamics, Arch. Rat. Mech. Anal., **177**(2005), 185–229.

[21] B. Guo and P. Zhu, Asymptotic behaviour of the solution to the system for a viscous reactive gas, J. Differential Equations, **155**(1999), 177–202.

[22] D. Hoff, Global well-posedness of the Cauchy problem for the Navier-Stokes equations of nonisentropic flow with discontinuous initial data, J. Differential Equations, **95**(1992), 33–74.

[23] D. Hoff, Discontinuous solutions of the Navier-Stokes equations for multidimensional heat conducting fluids, Arch. Rat. Mech. Anal., **139**(1997), 303–354.

[24] L. Hsiao and T. Luo, Large-time behavior of solutions for the outer pressure problem of a viscous heat-conductive one-dimensional real gas, Proc. Roy. Soc. Edinburgh A, **126**(1996), 1277–1296.

[25] N. Itaya, On the Cauchy problem for the system of fundamental equations describing the movement of compressible fluids, Kodai Math. Sem. Rep., **23**(1971), 60–120.

[26] S. Jiang, On initial-boundary value problems for a viscous, heat-conducting, one-dimensional real gas, J. Differential Equations, **110**(1994), 157–181.

[27] S. Jiang, On the asymoptotic behavior of the motion of a viscous, heat-conducting, one-dimensional real gas, Math. Z., **216**(1994), 317–336.

[28] S. Jiang, Global spherically symmetric solutions to the equations of a viscous polytropic ideal gas in an exterior domain, Comm. Math. Phys., **178**(1996), 339–374.

[29] S. Jiang, Large-time behavior of solutions to the equations of a viscous polytropic ideal gas, Ann. Mat. Pure Appl., **CLXXV** (1998), 253–275.

[30] D. Kassoy and J. Poland, The induction period of a thermal explosion in a gas between infinite parallel plates, Combust. Flame, **50**(1983), 259–274.

[31] S. Kawashima, T.Nishida, Global solutions to the initial boundary value problems for the equations of one-dimensional motion of viscous polytropic gases, J. Math. Kyoto Univ., **21**(1981), 825–837.

[32] B. Kawohl, Global existence of large solutions to initial-boundary value problems for a viscous, heat-conducting, one-dimensional real gas, J. Differential Equations, **58**(1985), 76–103.

[33] A.V. Kazhikhov, Sur la solubilité globale des problèmes monodimensionnels aux valeurs initiales-limitées pour les équations d'un gaz visqueux et calorifére, C. R. Acad. Sci. Paris Sér. A, **284**(1977), 317–320.

[34] A.V. Kazhikhov, To a theory of boundary value problems for equations of one-dimensional nonstationary motion of viscous heat-conduction gases, in Boundary Value Problems for Hydrodynamical Equations, No. 50, Institute of Hydrodynamics, Siberian Branch Acad. USSR, 1981, 37–62 (in Russian).

[35] A.V. Kazhikhov, Cauchy problem for viscous gas equations, Siberian Math. J., **23**(1982), 44–49.

[36] A.V. Kazhikhov and V.V. Shelukhin, Unique global solution with respect to time of initial-boundary value problems for one-dimensional equations of a viscous gas, J. Appl. Math. Mech., **41**(1977), 273–282.

[37] M. Lewicka and S.J. Watson, Temporal asymptotics for the pth power Newtonian fluid in one space dimension, Z. angew. Math. Phys., **54**(2003), 633–651.

[38] M. Lewicka and P.B. Mucha, On temporal asymptotics for the pth power viscous reactive gas, Nonlinear Anal., TMA, **57**(2004), 951–969

[39] T. Luo, On the outer pressure problem of a viscous heat-conducting one dimensional real gas, Acta Math. Appl. Sinica, **13**(1997), 251–264.

[40] A. Matsumura and T. Nishida, The initial boundary value problems for the equations of motion of compressible viscous and heat-conductive fluids, Proc. Japan Acad. Ser. A, **55**(1979), 337–342.

[41] A. Matsumura and T. Nishida, The initial boundary value problems for the equations of motion of compressible viscous and heat-conductive gases, J. Math. Kyoto Univ., **20**(1980), 67–104.

[42] A. Matsumuura and T. Nishida, Initial boundary value problems for the equations of motion of general fluids, in: V.R. Glowinski, J.L. Lions (Eds.), Computing Methods in Applied Science and Engineering, North-Holland, Amsterdam, 1982, pp. 389–406.

[43] A. Matsumura and T. Nishida, Initial boundary value problems for the equations of motion of compressible viscous and heat-conductive fluids, Comm. Math. Phys., **89**(1983), 445–464.

[44] D. Mihalas and B. Weibel-Mihalas, Foundation of radiation hydrodynamics, Oxford Univ. Press, New York, 1984.

[45] T. Nagasawa, On the one-dimensional motion of the polytropic ideal gas nonfixed on the boundary, J. Differential Equations, **65**(1986), 49–67.

[46] T. Nagasawa, Global asymptotics of the outer pressure problem with free boundary, Japan J. Appl. Math., **5**(1988), 205–224.

[47] M. Okada and S. Kawashima, On the equations of one-dimensional motion of compressible viscous fluids, J. Math. Kyoto Univ., **23**(1983), 55–71.

[48] G. Pomraning, The Equations of Radiation Hydrodynamics, Pergamon Press, 1973.

[49] Y. Qin, Global existence and asymptotic behavior of solutions to a nonlinear hyperbolic-parabolic coupled systems with arbitary initial data, Ph.D. Thesis, Fudan University, 1998.

[50] Y. Qin, Global existence and asymptotic behavior of solutions to a system of equations for a nonlinear one-dimensional viscous, heat-conducting real gas, Chin. Ann. Math., **20A**(1999), 343–354 (in Chinese).

[51] Y. Qin, Global existence and asymptotic behavior for the solutions to nonlinear viscous, heat-conductive, one-dimensional real gas, Adv. Math. Sci. Appl., **20**(2000), 119–148.

[52] Y. Qin, Global existence and asymptotic behavior for a viscous, heat-conductive, one-dimensional real gas with fixed and thermally insulated endpoints, Nonlinear Anal., TMA, **44**(2001), 413–441.

[53] Y. Qin, Global existence and asymptotic behavior of solution to the system in one-dimensional nonlinear thermoviscoelastcity, Quart. Appl. Math., **59**(2001), 113–142.

[54] Y. Qin, Global existence and asymptotic behavior for a viscous, heat-conductive, one-dimensional real gas with fixed and constant temperature boundary conditions, Adv. Differential Equations, **7**(2002), 129–154.

[55] Y. Qin, Exponential stability for a nonlinear one-dimensional heat-conductive viscous real gas, J. Math. Anal. Appl., **272**(2002), 507–535.

[56] Y. Qin, Global existence for the compressible Navier Stokes equations with Dirichlet boundary conditions, J. Henan Univ., **33**(2003), 1–8.

[57] Y. Qin, Universal attractor in H^4 for the nonlinear one-dimensional compressible Navier-Stokes equations, J. Differential Equations, **207**(2004), 21–72.

[58] Y. Qin, Exponential stability and maximal attractors for a one-dimensional nonlinear thermoviscoelasticity, IMA J. Appl. Math., **70**(2005), 509–526.

[59] Y. Qin, Nonlinear Parabolic-Hyperbolic Systems and Their Attractors, Vol. **184**, Operator Theory, Advances in PDEs, Basel, Boston-Berlin, Birkhäuser, 2008.

[60] Y. Qin and F. Hu, Global existence and exponential stability for a real viscous heat-conducting flow with shear viscosity, Nonlinear Anal., Real World Applications, **10**(2009), 298–313.

[61] Y. Qin and L. Huang, Regularity and exponential stability of the p th power Newtonian fluid in one space dimension, Math. Models and Methods Appl. Sci., **20**(2010), 589–610.

[62] Y. Qin and L. Huang, Global existence and exponential stability for the ph power viscous reactive gas, Nonlinear Anal., TMA, **73**(2010), 2800–2818.

[63] Y. Qin and L. Huang, On the $1D$ viscous reactive and radiative gas with the one-order Arrhenius kinetics, Preprint.

[64] Y. Qin, L. Huang and Z. Ma, Global existence and exponential stability in H^4 for the compressible Navier-Stokes equations, Comm. Pure Appl. Anal., **8**(2009), 1991–2012.

[65] Y. Qin, T. Ma, M.M. Cavalcanti and D. Andrade, Exponential stability in H^4 for the Navier-Stokes equations of viscous and heat-conductive fluid, Comm. Pure Appl. Anal., **4**(2005), 635–664.

[66] Y. Qin and J.E. Muñoz Rivera, Universal attractors for a nonlinear one-dimensional heat-conductive viscous real gas, Proc. Roy. Soc. Edinburgh A, **132**(2002), 685–709.

[67] Y. Qin and Jaime E.Muñoz Rivera, Global existence and exponential stability in one-dimensional nonlinear thermoelasticity with thermal memory, Nonlinear Anal., TMA, **51**(2002), 11–32.

[68] Y. Qin and J.E. Muñoz Rivera, Large-time behaviour of energy in elasticity, J. Elasticity, **66**(2002), 171–184.

[69] Y. Qin and J.E. Muñoz Rivera, Polynomial decay for the energy with an acoustic boundary condition, Appl. Math. Lett., **16**(2003), 249–256.

[70] Y. Qin and J.E. Muñoz Rivera, Exponential stability and universal attractors for the Navier-Stokes equations of compressible fluids between two horizontal parallel plates in $\mathbb{R}^3$, Appl. Numer. Math., **47**(2003), 209–235.

[71] Y. Qin and J.E. Muñoz Rivera, Global existence and exponential stability of solutions to thermoelastic equations of hyperbolic type, J. Elasticity, **5**(2005), 125–145.

[72] Y. Qin and S. Wen, Global existence of spherically symmetric solutions for nonlinear compressible Navier-Stokes equations, J. Math. Phys.,**49(2)**(2008), 023101, 25pp.

[73] Y. Qin, Y. Wu and F. Liu, On the Cauchy problem for one-dimensional compressible Navier-stokes equations, Portugaliae Mathematica, **64**(2007), 87–126.

[74] P. Secchi, On the motion of gaseous stars in the presence of radiation, Comm. Partial Differential Equations, 15(1990), 185–204.

[75] M. Umehara and A. Tani, Global solutions to the one-dimensional equations for a self-gravitating viscous radiative and reactive gas, J. Differential Equations, 234(2007), 439–463.

[76] D. Wang, The initial-boundary-value problem for the real viscous heat-conducting flow with shear viscosity, Proc. Roy. Soc. Edinburgh, 133A(2003), 968–986.

[77] F. Williams, Combustion Theory, Benjamin/Cummings, Merrlo Park, 1985.

[78] C. Xu and T. Yang, Local existence with physical vacuum boundary condition to Euler equations with damping, J. Differential Equations, 210(2005), 217–231.

[79] S. Yanagi, Existence of periodic solutions for a one-dimensional isentropic model system of compressible viscous gas, Nonlinear Anal., TMA, 46(2001), 279–298.

[80] H. Fujita-Yashima and R. Benabidallah, Unicité de la solution de l'équation monodimensionnelle ou à symétrie sphérique d'un gaz visqueux calorifére, Rendi del Circolo Mat. di Palermo, Ser II, 42(1993), 195–218.

[81] H. Fujita-Yashima and R. Benabidallah, Equation à symétrie sphérique d'un gaz visqueux et calorifére avec la surface libre, Ann. Mat. Pura Appl., 168(1995), 75–117.

[82] S. Zheng, Nonlinear Evolution Equations, Pitman Monographs and Surveys in Pure and Applied Mathematics, Vol. 133(2004), CRC Press, USA.

[83] S. Zheng and Y. Qin, Maximal attractor for the system of one-dimensional polytropic viscous ideal gas, Quart. Appl. Math., 59(2001), 579–599.

[84] S. Zheng and Y. Qin, Universal attractors for the Navier-Stokes equations of compressible and heat-conductive fluid in bounded annular domains in $\mathbb{R}^n$, Arch. Rat. Mech. Anal., 160(2001), 153–179.

[85] S. Zheng and W. Shen, Global solutions to the Cauchy problem of the equations of one-dimensional thermoviscoelasticity, J. Partial Differential Equations, 2(1989), 26–38.

[86] J. Zimmer, Global existence for a nonlinear system in thermoviscoelasticity with nonconvex energy, J. Math. Anal. Appl., 292(2004) 589–604.

Index

absolute temperature, 2
annular domains, 1

classical solutions, 3
compressible
   heat-conducting flow, 33
conservation of mass, 94

dissipative processes, 33

energy method, 3
exponential stability, 33, 93, 127
external force, 1

global existence, 1, 33, 93, 127

heat source, 1
Helmholtz free energy function, 128

ideal gas, 128
initial boundary value problem, 1, 34, 128

kinetic energy, 33

large-time behavior , 162
longitudinal velocity, 34

Navier-Stokes equations, 33, 93
Newtonian fluid, 33
non-autonomous compressible
   Navier-Stokes equations, 1

radiative gas, 127
real gas, 128

shear viscosity, 33
specific heat capacity, 2
spherically symmetric solutions, 1
Stefan-Boltzmann model, 127
stress tensor, 33
symmetric motion, 2

transverse velocity, 34

uniform estimates, 32, 163

viscosity coefficients, 2, 33
viscous polytropic ideal gas, 2